SIGNALING BY TOLL-LIKE RECEPTORS

METHODS IN SIGNAL TRANSDUCTION SERIES

Joseph Eichberg, Jr., Series Editor

Published Titles

Lipid Second Messengers, Suzanne G. Laychock and Ronald P. Rubin

G Proteins: Techniques of Analysis, David R. Manning

Signaling Through Cell Adhesion Molecules, Jun-Lin Guan

G Protein-Coupled Receptors, Tatsuya Haga and Gabriel Berstein

Calcium Signaling, James W. Putney, Jr.

G Protein-Coupled Receptors: Structure, Function, and Ligand Screening, Tatsuya Haga and Shigeki Takeda

Calcium Signaling, Second Edition, James W. Putney, Jr.

Analysis of Growth Factor Signaling in Embryos, Malcolm Whitman and Amy K. Sater

Signaling in the Retina, Steven J. Fliesler and Oleg G. Kisselev

SIGNALING BY TOLL-LIKE RECEPTORS

Edited by

Gregory W. Konat

CRC Press
Taylor & Francis Group
Boca Raton London New York

CRC Press is an imprint of the
Taylor & Francis Group, an **informa** business

CRC Press
Taylor & Francis Group
6000 Broken Sound Parkway NW, Suite 300
Boca Raton, FL 33487-2742

First issued in paperback 2019

© 2008 by Taylor & Francis Group, LLC
CRC Press is an imprint of Taylor & Francis Group, an Informa business

No claim to original U.S. Government works

ISBN-13: 978-1-4200-4318-1 (hbk)
ISBN-13: 978-0-367-38718-1 (pbk)

Library of Congress Cataloging-in-Publication Data

Signaling by toll-like receptors / editor, Gregory W. Konat.
    p. ; cm. -- (Methods in signal transduction series)
   Includes bibliographical references and index.
   ISBN 978-1-4200-4318-1 (hardcover : alk. paper) 1. Cellular signal
transduction. 2. Natural immunity. 3. Cell receptors. 4. Cellular immunity. I.
Konat, Gregory W. II. Series: Methods in signal transduction.
   [DNLM: 1. Toll-Like Receptors--physiology. 2. Signal
Transduction--physiology. QW 570 S5787 2008]

QP517.C45.S557 2008
571.7'4--dc22   2008012853

Visit the Taylor & Francis Web site at
http://www.taylorandfrancis.com

and the CRC Press Web site at
http://www.crcpress.com

Contents

Series Preface

The concept of signal transduction at the cellular level is now established as a cornerstone of the biological sciences. Cells sense and react to environmental cues by means of a vast panoply of signaling pathways and cascades. While the steady accretion of knowledge regarding signal transduction mechanisms is continuing to add layers of complexity, this greater depth of understanding has also provided remarkable insights into how healthy cells respond to extracellular and intracellular stimuli and how these responses can malfunction in many disease states.

Central to advances in unraveling signal transduction is the development of new methods and refinement of existing ones. Progress in the field relies upon an integrated approach that utilizes techniques drawn from cell and molecular biology, biochemistry, genetics, immunology and computational biology. The overall aim of this series is to collate and continually update the wealth of methodology now available for research into many aspects of signal transduction. Each volume is assembled by one or more editors who are leaders in their specialty. Their guiding principle is to recruit knowledgeable authors who will present procedures and protocols in a critical yet reader-friendly format. The goal is to assure that each volume will be of maximum practical value to a broad audience, including students, seasoned investigators and researchers who are new to the field.

The history of Toll-like receptors (TLRs) spans a mere 20 years, beginning with their discovery as products of the *Drosophila* Toll gene that play a role in dorsoventral patterning during development. Only several years later were TLRs recognized to play a key role in innate immunity, first in insects and subsequently in humans. During the past decade, this field has expanded at an extremely rapid rate. As a result, the transduction pathways through which members of this receptor family signal are becoming better defined, although our understanding is far from complete.

The chapters in this volume provide an overall view of our current knowledge of the role of TLRs and at the same time describe a variety of techniques that have been and are being utilized in investigations of TLR signaling mechanisms. These methods are presented in step-by-step, detailed protocols that are intended to be readily utilized by researchers. It is hoped that this compilation will prove to be of timely value to all investigators who are interested in this area.

Joseph Eichberg, Ph.D.
Advisory Editor for the Series

Preface

The discovery of Toll-like receptors (TLRs) a decade ago was seminal to our understanding of how innate immune cells recognize a variety of exogenous (i.e., microbial) as well as endogenous (i.e., host-derived) pathogenic molecules. Subsequently, the mechanisms by which these receptors trigger inflammatory responses that are appropriately tailored to the insult have been revealed. These studies provided a quantum leap that spurred the field of innate immunity into a renaissance after many years of neglect. As a result of this renewed research interest, it has become apparent that phylogenetically conserved innate immunity not only provides the first line of defense against invading microbes but also governs the response to tissue injury and controls the activity of adaptive immunity. Consequently, targeting TLRs offers promising new therapeutic strategies for the treatment of acute and chronic inflammatory conditions through the modification of immune responses. By extension, such strategies would be highly applicable to various autoimmune and allergic diseases, ischemic and traumatic conditions, and cancer.

The objective of *Signaling by Toll-Like Receptors* is to provide scientists at all levels of experience a combination of theoretical and practical information for understanding and studying TLRs. The ten chapters of this book introduce key concepts regarding TLR function and provide the first comprehensive compilation of practical procedures for a broad range of techniques that can be utilized for examining critical steps associated with the TLR-signaling process. The protocols describe step-by-step instructions on specific assays as well as critical comments necessary for their successful execution. The reader will find that the chapters often overlap in many areas. However, these inevitable repetitions occur in different contexts and are necessary for the integrity of individual chapters.

Chapter 1 describes approaches to evaluate cellular response to TLR stimulation. The signaling pathways triggered by TLR ligation ultimately converge on a few transcription factors such as nuclear factor kappa B and interferon regulatory factors, as well as on mitogen-activated protein kinases that in combination cooperate to induce the expression of proinflammatory genes. Consequently, monitoring activation of these factors provides reliable readouts of TLR stimulation. Chapter 2 addresses an important issue regarding the choice of ligands for such studies. Because both exogenous and endogenous ligands are frequently cross-contaminated, critical scrutiny of ligand purity is a sine qua non for the identification of their cognate receptors.

The following three chapters describe events along the intracellular signaling pathways. Chapter 3 deals with the regulation of TLR expression and identifies pivotal steps in the downstream signaling pathways emanating from TLRs (i.e., the formation and activation of the so-called signaling complexes). The ligation of TLRs also leads to their internalization and translocation into various subcellular compartments whereby the receptors are either degraded or recycled. This process is addressed in chapter 4. In chapter 5, a novel regulatory mechanism involving the Rho proteins is introduced. The Rho proteins are small GTPases that interact with

TLR-signaling pathways; however, the sites of these interactions seem to be cell and receptor dependent. Chapters 6 and 7 complement the previous chapters by describing recombinant stratagems to study interactions of TLRs with other signaling proteins and mechanisms that target the receptors to different cellular membranes.

The last three chapters address the interplay of TLR signaling with other pathophysiological processes. Chapter 8 examines cross talk between TLR and G protein–coupled receptor signaling in the context of leukocyte response to chemokine stimulation. Chapter 9 probes the role of TLRs in response to central nervous system infection, focusing on the effect of TLR signaling in regulating gap junction communication between glial cells. Finally, chapter 10 discusses the involvement of TLR signaling in the increased inflammatory reactivity that results from burn injury.

I hope the readers will find this book a user-friendly and helpful resource of theoretical and applied information. I also hope that this information will allow investigators to undertake their own studies to further expand this fascinating field of research.

Gregory W. Konat

About the Editor

Dr. Gregory W. Konat is a professor in the Department of Neurobiology and Anatomy, West Virginia University School of Medicine, Morgantown. He obtained his research training in biochemistry with a master's degree from the University of Warsaw and a doctorate from the University of Southern Denmark. Prior to joining the West Virginia University faculty, he worked at several institutions, namely, the Medical Research Center of the Polish Academy of Sciences in Warsaw, the Neurochemical Institute in Copenhagen, the University of Copenhagen, the University of Texas Medical School at Houston, and the Medical University of South Carolina in Charleston. Dr. Konat's initial research interests centered on central nervous system myelination and myelin disorders. In particular, his endeavors involved studies of the assembly of myelin membrane, the regulation of myelin gene expression, as well as the pathochemistry and immunology of multiple sclerosis. His current research emphasis has shifted toward inflammatory mechanisms in the central nervous system, especially oxidative damage and innate immune responses of glial cells.

Contributors

Derek W. Abbott
Department of Pathology
School of Medicine
Case Western Reserve University
Cleveland, OH

Anthony L. DeFranco
Department of Microbiology and
 Immunology
University of California, San Francisco
San Francisco, CA

Terje Espevik
Department of Cancer Research and
 Molecular Medicine
Norwegian University of Science and
 Technology (NTNU)
Trondheim, Norway

Øyvind Halaas
Department of Cancer Research and
 Molecular Medicine
Norwegian University of Science and
 Technology (NTNU)
Trondheim, Norway

Kenya Honda
Laboratory of Immune Regulation
Department of Microbiology and
 Immunology
Graduate School of Medicine
Osaka University
Osaka, Japan

Harald Husebye
Department of Cancer Research and
 Molecular Medicine
Norwegian University of Science and
 Technology (NTNU)
Trondheim, Norway

John H. Kehrl
Laboratory of Immunoregulation
National Institute of Allergy and
 Infectious Diseases
National Institutes of Health
Bethesda, MD

Tammy Kielian
Department of Neurobiology and
 Developmental Sciences
University of Arkansas for Medical
 Sciences
Little Rock, AR

Ulla G. Knaus
Department of Immunology, IMM-28
The Scripps Research Institute
La Jolla, CA

James A. Lederer
Department of Surgery
Brigham and Women's Hospital
Harvard Medical School
Boston, MA

Hyun-Ku Lee
Department of Immunology, IMM-12
The Scripps Research Institute
La Jolla, CA

Xiaoxia Li
Department of Immunology
Cleveland Clinic Foundation
Lerner Research Institute
Cleveland, OH

Adrian A. Maung
Department of Surgery
Brigham and Women's Hospital
Harvard Medical School
Boston, MA

Thomas J. Murphy
Department of Surgery
Brigham and Women's Hospital
Harvard Medical School
Boston, MA

Tadashi Nishiya
Department of Cellular Pharmacology
Hokkaido University Graduate School
  of Medicine
Sapporo, Japan

Hugh M. Paterson
Department of Surgery
Brigham and Women's Hospital
Harvard Medical School
Boston, MA

Monica Ruse
Department of Immunology, IMM-28
The Scripps Research Institute
La Jolla, CA

Chong-Shan Shi
Laboratory of Immunoregulation
National Institute of Allergy and
  Infectious Diseases
National Institutes of Health
Bethesda, MD

Harald Stenmark
Department of Cancer Research and
  Molecular Medicine

Norwegian University of Science and
  Technology (NTNU)
Trondheim, Norway

Kiyoshi Takeda
Laboratory of Immune Regulation
Department of Microbiology and
  Immunology
Graduate School of Medicine
Osaka University
Osaka, Japan

Peter S. Tobias
Department of Immunology, IMM-12
The Scripps Research Institute
La Jolla, CA

Min-Fu Tsan
Mid-Atlantic Regional Office (10R)
Office of Research Oversight
Department of Veterans Affairs
Washington, DC

Hui Xiao
Department of Immunology
Cleveland Clinic Foundation
Lerner Research Institute
Cleveland, OH

Masahiro Yamamoto
Laboratory of Immune Regulation
Department of Microbiology and
  Immunology
Graduate School of Medicine
Osaka University
Osaka, Japan

Abbreviations

4EX: extracellular domain of TLR4
AGA: 18-α-glycyrrhetinic acid
AKT: v-akt oncogene homologue; also termed "protein kinase B (PKB)"
AM: acetoxymethyl ester
AP-1: activator protein 1
APC: antigen-presenting cell
ATF2: activating transcription factor 2
BBB: blood-brain barrier
Bla: β-lactamase
BLP: bacterial lipopeptide
BMDM: bone marrow–derived macrophage
BSA: bovine serum albumin
cAMP: adenosine 3′, 5′-cyclic monophosphate
CCD: charge-coupled device
CD: cluster of differentiation
cDC: conventional dendritic cell
cDNA: complementary DNA
CFP: cyan fluorescent protein
CpG: cytosine linked to a guanine by a phosphate bond
CT: computed tomography
Cx: connexin
Cy: cyanine
DAMP: damage-associated molecular pattern
DAPI: 4′, 6-diamidino-2-phenylindole
DC: dendritic cell
DD: death domain
DHFR: dihydrofolate reductase
DMEM: Dulbecco's modified Eagle's medium
DMSO: dimethyl sulfoxide
dNTP: deoxyribronucleotide triphosphate
DOC: deoxycholate
ds: double stranded
DTT: dithiothreitol
EAE: experimental autoimmune encephalomyelitis
ECL: enhanced chemiluminescence
EDTA: ethylenediaminetetraacetic acid
EEA1: early endosome antigen 1
EGFP: enhanced green fluorescent protein
EGFR: epidermal growth factor receptor
ELISA: enzyme-linked immunosorbent assay
ER: endoplasmic reticulum

ERK: extracellular signal-regulated kinase
ESCRT: endosomal sorting complex required for transport
EU: endotoxin unit
FACS: fluorescence-activated cell sorting
FBS: fetal bovine serum
FCS: fetal calf serum
FRET: fluorescence resonance energy transfer
FSC: forward scatter
GCM: glial culture medium
GDI: guanine nucleotide dissociation inhibitor
GDP: guanosine diphosphate
GEF: guanine nucleotide exchange factor
GFAP: glial fibrillary acidic protein
GFP: green fluorescent protein
GJC: gap junction communication
GM-CSF: granulocyte-macrophage colony-stimulating factor
Gnai2: GTP-binding regulatory protein Gi alpha-2
gp96: glucose-regulated protein 96
GPCR: G protein–coupled receptor
GPI: glycosylphosphatidylinositol
GST: glutathione
GTP: guanosine triphosphate
GZA: glycyrrhizic acid
HBSS: Hank's buffered salt solution
HEK: human embryonic kidney
HEPES: 4-(2-hydroxyethyl)-1-piperazineethanesulfonic acid
HMGB1: high-mobility group box 1
HRP: horseradish peroxidase
Hrs: hepatocyte growth factor–regulated tyrosine kinase substrate
HSP: heat shock protein
ICP: intracranial pressure
ID: intermediate domain
IFN: interferon
IgG: immunoglobulin G
IκBα: inhibitor of NF-κB
IKK: inhibitor of NF-κB kinase
IL: interleukin
IL-1R: interleukin-1 receptor
IRAK: IL-1 receptor–associated kinase
IRES: internal ribosome entry site
IRF: interferon regulatory factor
JNK: c-Jun N-terminal kinase
KO: knockout
LAL: limulus amebocyte lysate
LD50: 50 percent lethal dose
L-LME: L-leucine methyl ester

LPS: lipopolysaccharide
LRR: leucine-rich repeat
LTA: lipoteichoic acid
LY: Lucifer yellow
Mal: MyD88 adaptor-like
MAPK: mitogen-activated protein kinase
MBP: myelin basic protein
M-CSF: macrophage colony–stimulating factor
MD-2: myeloid differentiation protein 2
MEKK3: MAPK kinase kinase 3
MHC: major histocompatibility complex
MKK6: MAPK kinase 6
mmLDL: minimally modified or oxidized low-density lipoprotein
MOF: multiple organ failure
MRI: magnetic resonance imaging
MS: multiple sclerosis
MyD88: myeloid differentiation factor 88
NEMO: NF-κB essential modulator
NF-κB: nuclear factor κB
NO: nitric oxide
NOD: nucleotide oligomerization domain
OCT: optimal cutting temperature medium
ODN: oligodeoxynucleotide
OPI: oxalacetic acid, pyruvate, and insulin
PAGE: polyacrylamide gel electrophoresis
Pak1: p21-activated kinase 1
PAMP: pathogen-associated molecular pattern
PBD: Pak1 binding domain
PBS: phosphate buffered saline
PCA: protein complementation assay
PCR: polymerase chain reaction
pDCs: plasmacytoid dendritic cells
PE: phycoerythrin
PEG: polyethylene glycol
Pellino: Pelle-interacting protein
PGN: peptidoglycan
PID: Pak1 inhibitory domain
PKC: protein kinase C
PMSF: phenylmethylsulfonyl fluoride
PRR: pattern recognition receptor
Pyk2: proline-rich tyrosine kinase 2
qRT-PCR: quantitative real-time reverse transcriptase polymerase chain reaction
RAGE: receptor for advanced glycation end products
RGS: regulators of G protein signaling
***S. aureus*:** *Staphylococcus aureus*
SIGIRR: single immunoglobulin IL-1 receptor-related molecule

siRNA: small interfering RNA
SIRS: systemic inflammatory response syndrome
SP-A: surfactant protein-A
ss: single stranded
SSC: side scatter
STAT: signal transducer and activator of transcription
TAB: TAK1 binding protein
TAK1: TGFβ-activated kinase 1
TBK1: TANK-binding kinase 1
TBSA: total body surface area
Tf-R: transferrin receptor
TGF: transforming growth factor
TICAM: TIR domain–containing adaptor molecule
TIR: Toll or interleukin-1 receptor homology domain
TIRAP: TIR domain–containing adaptor protein
TLR: Toll-like receptor
TM-CP: transmembrane and cytoplasmic domain
TNF: tumor necrosis factor
Tollip: Toll inhibitory protein
TRAF: TNF receptor–associated factor
TRAM: TRIF-related adaptor molecule
TRIF: TIR domain–containing adaptor inducing interferon-β
Ub: ubiquitin
UbR: nonfunctional ubiquitin
UbRGG: functional wild-type ubiquitin
UIM: ubiquitin interacting motif
WT: wild type
YFP: yellow fluorescent protein

List of Protocols

1 Assessing the Response of Cells to TLR Stimulation

Kiyoshi Takeda, Masahiro Yamamoto, and Kenya Honda

CONTENTS

1.1 INTRODUCTION

1.1.1 TOLL-LIKE RECEPTORS

Toll was first identified as an essential receptor for dorsoventral patterning in the developing embryo in *Drosophila*. In 1996, it was discovered that *Drosophila* Toll is involved in antifungal responses in the adult fly.[1] Since then, the role of Toll receptors in the innate immune response has been studied intensively, not only in insects but also in mammals. This has led to the identification of Toll receptors in mammals, named Toll-like receptors (TLRs). The first mammalian homolog of *Drosophila* Toll was identified in 1997 as hToll (now termed TLR4).[2] Subsequent studies have identified several proteins that are structurally related to TLR4. The TLR family now comprises ten members in humans (TLR1–TLR10) and twelve in the mouse (TLR1–TLR9 and TLR11–TLR13). The cytoplasmic portion of Toll-like receptors shows high similarity to that of the interleukin-1 (IL-1) receptor family, and is known as the Toll/IL-1 receptor (TIR) domain. Despite this similarity, the extracellular portions of both receptors are structurally unrelated. The IL-1 receptor is characterized by the presence of an Ig-like domain, whereas Toll-like receptors bear leucine-rich repeats in the extracellular domain.

The critical involvement of TLRs in the recognition of microorganisms has now been established. Each TLR recognizes specific patterns of microbial components (Table 1.1).[3] Recognition of microbial components by TLRs triggers activation of innate immunity through induction of gene expression. Furthermore, TLR-mediated activation of innate immunity has been shown to be mandatory for the development of antigen-specific adaptive immunity.[4]

1.1.2 SUBCELLULAR LOCALIZATION OF TLRs

TLRs are type I transmembrane molecules, harboring extracellular, transmembrane, and intracellular portions. However, individual TLRs are differentially distributed within the cell. TLR1, TLR2, TLR4, and TLR6 are expressed on the cell surface, as demonstrated by positive staining of the cell surface by specific antibodies. In contrast, TLR3, TLR7, TLR8, and TLR9 have been shown to be expressed in intracellular compartments such as endosomes.[5–8] TLR3-, TLR7-, and TLR9-mediated recognition of their ligands has been shown to require endosomal maturation.[9–12] Nonspecific uptake by endosomes of the TLR9 ligand CpG (cytosine linked to a guanine by a phosphate bond) DNA leads to recruitment of TLR9 from the endoplasmic reticulum.[8,12] It can be hypothesized that, in the case of bacterial infection,

TABLE 1.1
TLRs and Their Ligands

TLR	Ligands
TLR1	Triacyl lipopeptides
TLR2	Peptidoglycan lipopeptides, lipoteichoic acid, lipoarabinomannan, glycosylphosphatidylinositol (GPI) anchors, phenolsoluble modulin, zymosan, and glycolipids
TLR3	dsRNA
TLR4	Lipopolysaccharide, taxol, Rous sarcoma virus (RSV) fusion protein, mouse mammary tumor virus (MMTV) envelope protein, and endogenous ligand (heat shock proteins [HSPs], fibronectin, and hyaluronic acid)
TLR5	Flagellin
TLR6	Diacyl lipopeptides
TLR7	ssRNA, imidazoquinolines
TLR8	ssRNA, imidazoquinolines
TLR9	CpG DNA
TLR11	Profilin

following the phagocytosis of bacteria by macrophages and dendritic cells (DCs) and their degradation of bacteria in phagosome-lysosome or endosome-lysosome, CpG DNA is exposed and recognized by TLR9. In the case of viral infection, receptor-mediated virus entry is usually directed into the cytoplasm, but occasionally the virus is passed into the endosomal compartment, which results in degradation of viral particles, leading to exposure of TLR ligands such as dsRNA (double-stranded ribonucleic acid; TLR3 ligand), ssRNA (single-stranded ribonucleic acid; TLR7 ligand), and CpG DNA (TLR9 ligand). Even TLR2, which is expressed on the cell surface, is recruited to the phagosomal compartment of macrophages after exposure to zymosan.[13]

1.1.3 SIGNALING PATHWAYS THROUGH TLRs

Stimulation of TLRs by microbial components triggers activation of signaling pathways and thereby induces expression of several genes that are involved in immune responses. Microbial recognition of TLRs facilitates their dimerization. TLR2 is shown to form a heterophilic dimer with TLR1 or TLR6, but in other cases TLRs form homodimers.[14] Dimerization of TLRs triggers activation of signaling pathways that originate from the conserved cytoplasmic TIR domain. In the signaling pathways downstream of the TIR domain, a TIR domain–containing adaptor, myeloid differentiation factor 88 (MyD88), was first shown to be essential for induction of inflammatory cytokines such as tumor necrosis factor-α (TNF-α) and IL-12 through all TLRs except TLR3.[15,16] However, activation of specific TLRs leads to slightly different patterns of gene expression profiles. For example, activation of TLR3 and TLR4 signaling pathways results in induction of type I interferons (IFNs), but activation of TLR2- and TLR5-mediated pathways does not.[17,18] TLR7, TLR8, and TLR9 signaling pathways in a special subset of DCs (viz., plasmacytoid DCs) lead to induction of type I IFNs through mechanisms distinct from TLR3- and TLR4-mediated induction.[19–21] Thus, individual TLR signaling pathways are divergent, though MyD88 is common to all TLRs. It has now become clear that there are MyD88-dependent and MyD88-independent pathways.

1.1.4 MyD88-DEPENDENT PATHWAYS

As its name suggests, in the MyD88-dependent pathway MyD88 plays a crucial role. MyD88-deficient mice do not produce inflammatory cytokines such as TNF-α and IL-12p40 in response to any TLR ligands except the TLR3 ligand.[15,16] MyD88 harbors a C-terminal TIR domain and an N-terminal death domain, and associates with the TIR domain of TLRs. On stimulation, MyD88 recruits IRAK-4 (IL-1 receptor-associated kinase 4) to TLRs through an interaction of the death domains of both molecules, and facilitates IRAK-4-mediated phosphorylation of IRAK-1. Both IRAK-1 and TRAF6 (TNF receptor-associated factor 6) are polyubiquitinated in response to TLR ligands. Ubiquitin chains formed on IRAK-1 are linked through lysine 48 (K48) of ubiquitin, which mediates proteasome-dependent protein destruction. On stimulation, degradation of IRAK-1 is observed in the MyD88-dependent pathway. In contrast, ubiquitin chains on TRAF6 are linked through lysine 63 (K63), which was originally shown to be implicated in biological processes such as the stress response and DNA repair rather than protein destruction. Although the E3 ubiquitin ligases of IRAK-1 have not been identified, TRAF6 contains a RING finger domain in the N-terminal portion harboring E3 ubiquitin ligase activity, and mediates auto-ubiquitination. The K63-linked ubiquitination of TRAF6 is shown to further stimulate activation of the MAP3K (mitogen-activated protein kinase 3) family member, transforming growth factor-B-activated kinase 1 (TAK1) in concert with TAK1-associated molecules such as TAB2 and TAB3 (TAK1 binding proteins). In addition, a member of the ubiquitin-conjugating (Ubc) proteins, Ubc13, has been identified as the E2 ubiquitin-conjugating enzyme for K63-linked polyubiquitination. Since mice deficient in either TAK1 or Ubc13 exhibit early embryonic death, the relevance of these molecules was examined using conditionally ablated mice in which TLR-mediated immune responses were severely impaired.[22,23] Activation of TAK1 leads to two distinct transcription factors, NF-κB (nuclear factor κB) and AP-1 (activator protein 1). Although members of the inhibitory protein IκB family sequester NF-κB in the cytoplasm of unstimulated cells, stimulus-dependent IκB phosphorylation by the IκB kinase (IKK) complex and degradation by the K48-linked ubiquitin-proteasome pathway permit translocation of NF-κB to the nucleus. In contrast, AP-1 is activated by MAP kinases such as c-Jun N-terminal kinase (JNK) and p38. Activation of AP-1 and NF-κB leads to the induction of several genes such as those encoding proinflammatory cytokines (Figure 1.1). In addition to AP-1 and NF-κB, several of the interferon regulatory factor (IRF) family of transcription factors are activated by the MyD88-dependent pathways and contribute to the specific gene expression programs induced by TLRs. IRF-5 and IRF-7 interact directly with MyD88 and regulate the TLR-dependent induction of inflammatory cytokines and type I IFNs, respectively.[24–27] IRF-4 interacts with MyD88 and acts as a negative regulator of MyD88-IRF-5-mediated gene induction.[28] IRF-1 induced by IFN-α also forms a complex with MyD88, and is activated by MyD88 to migrate rapidly into the nucleus and mediate efficient induction of IFN-β, inducible nitric oxide synthase, and IL-12 p35.[29] IRF-8 (also known as ICSBP) interacts with TRAF6 and regulates the production of inflammatory mediators.[30]

A database search for molecules that are structurally related to MyD88 led to the identification of the second TIR domain–containing molecule, TIRAP (TIR domain–containing adaptor protein)/Mal (MyD88 adaptor-like).[31,32] Similar to MyD88-deficient macrophages, TIRAP/Mal–deficient macrophages show impaired inflammatory cytokine production in response to TLR4 and TLR2 ligands.[33,34] However, TIRAP/Mal–deficient mice are not impaired in their response to TLR3, TLR5, TLR7, or TLR9 ligands. Thus, TIRAP/Mal has been shown to selectively mediate the MyD88-dependent signaling pathway via TLR2 and TLR4.

A special subset of DCs have a morphology like that of plasma cells in the resting state but, when activated (or matured), have a DC-like morphology.[35] According to their morphology, these cells are called plasmacytoid dendritic cells (pDCs). pDCs produce a large amount of IFN-α/β when infected with viruses, and they are also called natural interferon-producing cells. In pDCs, activation of TLR7 or TLR9 results in activation of a unique MyD88-dependent pathway leading to induction of IFN-α/β.[21,36] In this pathway, MyD88 interacts directly with IRF-7, which is an essential transcription factor for IFN-α/β induction.[25–27] Since pDCs from mice

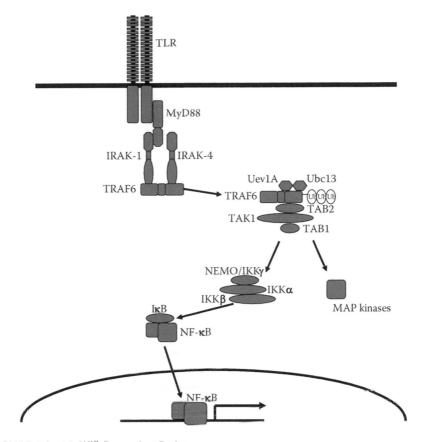

FIGURE 1.1 MyD88-Dependent Pathway

Note: Stimulation of TLRs triggers association with MyD88. MyD88, in turn, recruits IRAK-4 to the receptor complex, which allows association of IRAK-1 with MyD88. IRAK-4 induces phosphorylation of IRAK-1. TRAF6 is recruited to the receptor complex by association with phosphorylated IRAK-1. Phosphorylated IRAK-1 and TRAF6 dissociate from the receptor and form a complex with TAK1, TAB1, and TAB2 at the membrane, which induces phosphorylation of TAB2 and TAK1. IRAK-1 is degraded at the membrane, and the remaining complex comprising TRAF6, TAK1, TAB1, and TAB2 translocates to the cytosol, thereby associating with the ubiquitin ligases Ubc13 and Uev1A. This leads to ubiquitination of TRAF6, which induces activation of TAK1. TAK1, in turn, phosphorylates both the IKK complex (comprising IKKα, IKKβ, and NEMO/IKKγ) and MAP kinases. The IKK complex phosphorylates IκB, leading to its ubiquitination and subsequent degradation. This allows NF-κB to translocate into the nucleus and induce its target genes.

deficient in IRAK-1, IRAK-4, or IKKa show impaired induction of IFN-α/β in response to TLR7 or TLR9 stimulation, these molecules are implicated in the activation of IRF-7[37,38] (Figure 1.2). TLR9 is also expressed in other types of DCs such as conventional DCs. In conventional DCs, TLR9 stimulation does not lead to induction of type I IFNs despite robust induction of IL-12p40. The different induction mechanisms of the TLR9 ligand have been shown to be due to the sustained retention of the TLR9 ligand in the endosomal compartment in pDCs.[39]

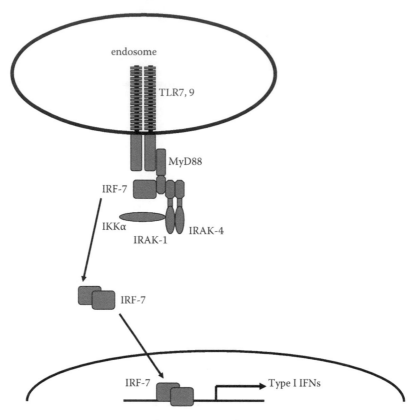

FIGURE 1.2   TLR Pathway in Plasmacytoid DCs (pDCs)

Note: In plasmacytoid DCs, stimulation of TLR7 and TLR9 leads to induction of type I IFNs in an MyD88-dependent manner. MyD88 associates with TLR7 and TLR9, which recognize nucleic acid-like ligands in the endosomal compartment. MyD88 also directly associates with IRF-7. TLR-dependent activation of IRF-7 is mediated by IRAK-4, IRAK-1, and IKKα. Activated IRF-7 translocates into the nucleus, where it induces expression of type I IFN genes.

1.1.5   MyD88-INDEPENDENT OR TRIF-DEPENDENT PATHWAYS

In MyD88-deficient macrophages, TLR4 ligand–induced production of inflammatory cytokines is not observed; however, activation of NF-κB is seen with delayed kinetics.[15] This indicates that although TLR4-mediated production of inflammatory cytokines completely depends on the MyD88-dependent pathway, a MyD88-independent component exists in TLR4 signaling. Indeed, TLR4 stimulation leads to activation of the transcription factor IRF-3, as well as activation of the late phase of NF-κB in a MyD88-independent manner.[40] TLR4-induced activation of IRF-3 leads to production of IFN-β. IFN-β in turn activates STAT1 (signal transducer and activator of transcription 1) and induces several IFN-inducible genes.[17,18] TLR3 stimulation also leads to activation of IRF-3 and thereby induces IFN-β in a MyD88-independent manner. TLR3 and TLR4 utilize the MyD88-independent component to induce IFN-β.

TIR domain–containing molecules, such as MyD88 and TIRAP/Mal, play important roles in the MyD88-dependent pathway. Similarly, other TIR domain–containing molecules regulate the MyD88-independent pathway. A database search led to the identification of a third TIR domain–containing adaptor, TIR domain–containing adaptor inducing IFN-β (TRIF).[41] This molecule was also identified as a TLR3-associated molecule by two-hybrid screening and was named TIR domain–containing adaptor molecule (TICAM-1).[42] The physiological role of TRIF–TICAM-1 has been demonstrated by the generation of TRIF-mutant mice. TRIF-deficient mice generated by gene targeting show no activation of IRF-3 and show impaired expression of IFN-β and IFN-inducible genes in response to TLR3 and TLR4 ligands.[43] Another mouse strain with a mutated *Trif* gene generated by random germline mutagenesis is also defective in TLR3- and TLR4-mediated induction of IFN-β and IFN-inducible genes.[44] Thus, TRIF has been demonstrated to be essential for the TLR3- and TLR4-mediated MyD88-independent pathway.

Further database searches led to the identification of a fourth TIR domain–containing adaptor, TRAM (TRIF-related adaptor molecules; also known as TICAM-2).[45,46] Studies of TRAM-deficient mice and RNA interference PE2 (RNAi-) mediated knockdown of TRAM expression show that TRAM is involved in TLR4-, but not TLR3-, mediated activation of IRF-3 and induction of IFN-β and IFN-inducible genes.[45,46] Thus, TRAM is essential for the TLR4-mediated MyD88-independent or TRIF-dependent pathway.

In TRIF- and TRAM-deficient mice, inflammatory cytokine production induced by TLR2, TLR7, and TLR9 ligands is observed. In addition, TLR4 ligand–mediated phosphorylation of IRAK-1 is normally induced.[43,46] These findings indicate that TLR-mediated activation of the MyD88-dependent pathway is not impaired in these mice. However, TLR4 ligand–induced inflammatory cytokine production is not observed in TRIF- and TRAM-deficient mice. These findings indicate that activation of both the MyD88-dependent and the MyD88-independent or TRIF-dependent components is required for TLR4-induced inflammatory cytokine production, but the mechanisms remain unknown.

Noncanonical IKKs, TBK1 and IKKi/IKKε, have been shown to mediate IRF-3 activation.[47] Introduction of TBK1 or IKKi/IKKε, but not IKKβ, results in phosphorylation and nuclear translocation of IRF-3. RNAi-mediated reduction of TBK1 or IKKi/IKKε expression leads to impaired induction of IFN-β in response to viruses and dsRNA.[47,48] Embryonic fibroblast cells from TBK1-deficient mice show impaired activation of IRF-3 and expression of IFN-β and IFN-inducible genes in response to TLR3 and TLR4 ligands.[49–51] In contrast, embryonic fibroblast cells from IKKi/IKKε-deficient mice are not defective in their response to TLR3 and TLR4 ligands.[50] However, TLR3-mediated activation of IRF-3 and expression of IFN-β and IFN-inducible genes are almost completely abolished in embryonic fibroblast cells lacking both TBK1 and IKKi/IKKε. Thus, TBK1 and IKKi/IKKε are critical regulators of IRF-3 activation in the MyD88-independent pathway.

TRIF interacts with receptor-interacting protein 1 (RIP1), which leads to TRIF-dependent NF-κB activation.[52] Embryonic fibroblast cells from RIP1-deficient mice show impaired NF-κB activation in response to the TLR3 ligand. Thus, RIP1 is

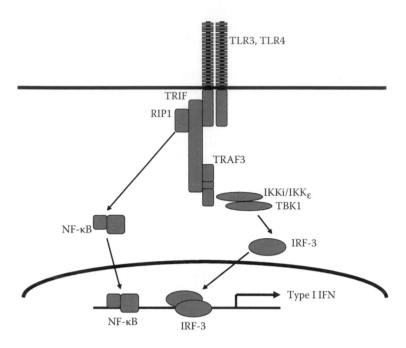

FIGURE 1.3   MyD88-Independent or TRIF-Dependent Pathway

Note: In TLR3- and TLR4-mediated signaling pathways, activation of IRF-3 and induction of IFN-β are observed in a MyD88-independent manner. TRIF is essential for the MyD88-independent pathway. TRAF3 links TRIF and nontypical IKKs (IKK*i*/IKKε and TBK1). IKK*i*/IKKε and TBK1 mediate activation of IRF-3 downstream of TRIF. TRIF also associates with RIP1, which mediates activation of NF-κB.

responsible for NF-κB activation via TRIF. TRIF also interacts with TRAF3, which bridges to TBK1 and IKK*i*/IKKε[53,54] (Figure 1.3).

1.2   MACROPHAGES

TLRs are expressed in several types of cells involved in innate immune responses, and induce gene expression. Among these cells, macrophages and DCs have been most intensively studied. These cell populations are heterogeneous and have several subsets. First, methods to isolate macrophages will be described. Macrophages residing in different tissues show distinct phenotypes in terms of surface expression of molecules such as activation markers and chemokine receptors as well as functional capacities such as phagocytic activity.[55] Thus, macrophages from different sites have distinct names; for example, macrophages residing in the liver are called Kupffer cells and cells residing in the brain are microglia cells. The differing properties of macrophages from different tissues are dependent on the microenvironment and the maturation stage of the cells. Macrophages are activated by microorganisms; therefore, macrophages residing in the sterile (noninflammatory) and nonsterile (inflammatory) sites are functionally different. Macrophages residing in sterile or noninflammatory sites are often referred to as "resident macrophages" and are relatively quiescent immunologically, having low oxygen consumption, low levels of major histocompatibility complex (MHC) class II expression, and low cytokine production. However, resident macrophages are highly phagocytic and chemotaxic, and retain some proliferative capacity. Macrophages from nonsterile (inflammatory) sites are often referred to as "activated" or "primed" macrophages, show an increased capacity to produce cytokines, and express several activation markers such as MHC class II.

1.2.1 Isolation of Peritoneal Macrophages

In the peritoneal cavity of mice, a few (about 1×10^5) cells are present, and most of these are macrophages expressing CD11b or F4/80. These resident macrophages can be isolated by washing the peritoneal cavity with 4 ml of Hank's buffered salt solution (HBSS) or phosphate buffered saline (PBS). As there are only a few resident macrophages in the peritoneal cavity, it is difficult to perform biochemical and molecular analyses of the effect of TLR stimulation. To increase the number of macrophages and even isolate "activated" or "primed" macrophages, peritoneal macrophages are often obtained after intraperitoneal injection of thioglycollate broth. Thioglycollate broth is a medium originally designed to test the aerotolerance of bacteria. Along with nutrients to support bacterial growth, it contains sodium thioglycollate, thioglycollic acid, L-cystine, methylene blue, and 0.05% agar. The sodium thioglycollate, thioglycollic acid, and L-cystine reduce the amount of oxygen in water. Methylene blue is an indicator that is colorless in an anaerobic environment and greenish-blue in the presence of oxygen. The agar helps to retard oxygen diffusion and to maintain the stratification of organisms growing in different layers of the broth.

Intraperitoneal injection of thioglycollate broth induces peritonitis in mice. During thioglycollate-elicited peritonitis, many inflammatory cells infiltrate the peritoneal cavity. In the very early phase after injection (within 6 hours), the major component of the infiltrating cell populations is neutrophils. After that time point, mainly macrophages are recruited into the peritoneal cavity, and at 3 days after injection, about 1×10^7 cells have been recruited. After this time, lymphocytes are increasingly recruited into the cavity. Thus, macrophages can be effectively isolated 3 days after injection of thioglycollate broth. At this time, many exudate cells are macrophages, as characterized by expression of CD11b or F4/80, and exhibit the properties of "activated" or "primed" macrophages.

Protocol 1.1: Isolation of Thioglycollate-Elicited Peritoneal Macrophages

1. Inject mice intraperitoneally with 2 ml of 4% thioglycollate broth (Sigma-Aldrich, St. Louis, Missouri).
2. After sacrifice or anesthetization, inject the mice intraperitoneally with 5 ml of HBSS or PBS at 3 days after thioglycollate injection.
3. Collect the injected HBSS or PBS, and transfer to a 15 ml culture tube. The collected buffer contains many peritoneal exudate cells.

4. Centrifuge cell suspension at 1,200 rpm for 5 minutes.
5. Suspend pellets with culture media (RPMI1640 containing 10% fetal calf serum [FCS] and 100 mM 2-mercaptoethanol).
6. Seed cells onto culture dish. (Size of dish depends on experiments. For example, 1×10^6 cells fit into a 6 cm dish.)
7. Incubate the cells for 2 hours at 37 °C, and wash them three times with HBSS. The remaining adherent cells are used as peritoneal macrophages for experiments.

1.2.2 ISOLATION OF BONE MARROW–DERIVED MACROPHAGES

Development of the macrophage lineage of cells from bone marrow progenitors is regulated by macrophage colony-stimulating factor (M-CSF), which is constitutively produced by many cell types. Accordingly, macrophages can be generated by culturing bone marrow cells with M-CSF. These bone marrow–derived macrophages are frequently used for experiments, since they are not affected by the microenvironment in which mice are housed.

Protocol 1.2: Preparation of Bone Marrow–Derived Macrophages

1. Take femora and tibias from mice. About 5×10^6 cells can be isolated from a mouse.
2. Sterilize the bones in 70% ethanol for 1 minute, and wash them with HBSS or PBS.
3. Cut the middle portion of the bones, and flush out bone marrow cells with 1 ml of HBSS or PBS in a syringe with a 26 gauge needle.
4. Pass the cells through 100 μm diameter nylon mesh (cell strainer from BD Biosciences, San Jose, CA), and centrifuge them at 1,200 rpm for 5 minutes.
5. Resuspend the cells in RPMI1640 medium supplemented with 10% FCS and 100 mM 2-mercaptoethanol.
6. Determine the cell number; adjust cell concentration to 1×10^6 cells/ml in RPMI1640 medium supplemented with 10% FCS, 100 mM 2-mercaptoethanol, and 10 ng/ml M-CSF (GenzymeTechne, Minneapolis, MN); and seed the cells onto a 10 cm dish.
7. Culture the cells at 37 °C under 5% CO_2.
8. Change the culture medium every 2 days. After 6–8 days, almost all adherent cells are positive for expression of F4/80 or CD11b and are used as macrophages.

1.3   DENDRITIC CELLS

The main function of DCs is to process engulfed proteins and present the protein-derived antigens on their surface together with MHC molecules to naïve T cells.[56] DCs are present in small quantities in mucosal tissues, such as the skin (where they are often called Langerhans cells), lungs, and intestines. They can also be found in an immature state in the blood. Once activated, they migrate to the lymphoid tissues, where they interact with naïve T cells to initiate the adaptive immune response. These DCs are also heterogeneous, and are divided into two main subsets: conventional DCs (cDCs) and plasmacytoid DCs (pDCs).[57] This subdivision was originally made in humans[58] and subsequently extended to the mouse.[59–61] Recent evidence indicates that plasmacytoid DCs do not have the property of antigen presentation; rather, they mainly produce type I IFNs in response to TLR stimulation or viral infection. In addition, plasmacytoid DCs activate other DCs, including those involved in cross-presentation.[62] Unlike conventional DCs, mouse plasmacytoid DCs express CD45RA and low levels of CD11c.

Conventional DCs are categorized into two subgroups: blood-derived DCs and tissue-derived DCs. Blood-derived DCs are present in the spleen and lymph nodes, and are further divided into three subsets based on surface expression of CD4 and CD8a: CD8+ DCs, CD4+ DCs, and CD4−CD8− DCs. Tissue-derived DCs comprise two subsets: the Langerhans cells and the dermal-interstitial DCs. These subsets express CD205,[63] though this marker is also expressed on CD8+ DCs. Tissue-derived DCs are CD8a^low CD11b+, while CD8+ DCs are CD8+CD11b−, and thus these cells can be distinguished. Dermal DCs of the skin or the interstitial DCs of other tissues are largely CD11b+CD205+CD8−. They differ from Langerhans cells in that they do not contain langelin granules, and they express lower levels of CD205 than Langerhans cells. Langerhans cells reside in the skin epithelia, where they detect invading pathogens and initiate immune responses. Although the precise role of CD4+ DCs and CD4−CD8− DCs is unclear, CD8+ DCs are well documented as being critical for T cell–mediated immunity, especially by producing IL-12. CD8+ DCs also mediate cross-presentation.

1.3.1   ISOLATION OF SPLENIC DENDRITIC CELLS

DCs residing in the spleen can be isolated using collagenase and ethylenediamine-tetraacetic acid (EDTA) to detach DCs from the interstitium.

Protocol 1.3: Isolation of DCs from Spleen

1. Dissect the spleen from a mouse, and cut it into small fragments.
2. Incubate the spleen with RPMI 1640 medium containing 400 U/ml collage-nase (Wako Pure Chemical Industries, Osaka, Japan) and 15 µg/ml DNase (Sigma-Aldrich, St. Louis, Missouri) at 37 °C for 30 minutes.

3. Add EDTA (final concentration 5 mM) for the last 5 minutes. This treatment allows DCs to detach from the interstitial connective tissues.
4. Prepare single-cell suspension by grinding down the spleen with slide glass.
5. Centrifuge cells at 1,200 rpm for 5 minutes.
6. Suspend pellets with 2 ml of lysis buffer (0.15 M NH_4Cl, 1 mM $KHCO_3$, and 0.1 mM EDTA [pH 7.2]) to lyse red blood cells.
7. Incubate for 5 minutes at room temperature, then add 10 ml of ice-cold HBSS.
8. Centrifuge cells at 1,200 rpm for 5 minutes.
9. Suspend pellets with PBS supplemented with 1% FCS and 2 mM EDTA.
10. Add 50 μl of CD11c-microbeads (Miltenyi Biotec, Bergisch Gladbach, Germany) to 500 μl of cell suspension (less than 1×10^7 cells/ml), and incubate at 4 °C for 20 minutes.
11. Purify CD11c$^+$ cells by MACS (Miltenyi Biotec, Bergisch Gladbach, Germany).

Enriched cells contain > 90% CD11c$^+$ cells. About 5×10^6 CD11c$^+$ cells can be isolated from a spleen. These cells are used as splenic DCs or are further purified for isolation of CD8$^+$ or CD8$^-$ DCs using an FCS sorting technique or the MACS system.

1.3.2 ISOLATION OF BONE MARROW–DERIVED DENDRITIC CELLS

DC progenitors in the bone marrow develop into DC precursors in the presence of granulocyte-macrophage colony-stimulating factor (GM-CSF), and circulate in the blood.[64] These DC precursors reside in the tissues, where they become immature DCs with a high capacity to engulf antigens. In addition to GM-CSF, Fms-like tyrosine kinase 3 (Flt3) ligand induces the development of DCs, but Flt3 ligand–induced DCs are functionally distinct from GM-CSF-induced DCs.[65] Bone marrow–derived DCs cultured in the presence of Flt3 ligand contain CD11c^{intermediate}CD11b$^-$B220$^+$ cells, which are phenotypically related to pDCs and produce large amounts of type I IFNs.[66,67]

Protocol 1.4: Isolation of Bone Marrow–Derived DCs

1. Isolate bone marrow cells as described in Protocol 1.2.
2. Determine the cell number, and suspend the cells (1×10^6 cells/ml) in RPMI1640 medium supplemented with 10% FCS, 100 mM 2-mercaptoethanol, and 10 ng/ml GM-CSF (PeproTech, Rocky Hill, NJ).
3. Culture the cells at 37 °C under 5% CO_2.
4. Change the culture medium every 2 days. After 6 days, the cells are used as DCs in experiments.

To prepare Flt3 ligand–induced DCs, bone marrow–derived DCs are cultured in RPMI 1640 medium supplemented with 10% FCS, 100 µM 2-ME, and 100 ng/ml human Flt3 ligand (PeproTech, Rocky Hill, NJ). The CD11c^{intermediate}CD11b⁻B220⁺ subset, which represents ~1% of the total bone marrow cells before culture, increases to about 15% of the total cultured cells at day 5, peaks on day 10 at 45%, and then decreases to 21% by day 15 of culture.

1.4   DETECTING THE ACTIVITY OF TLR-SIGNALING PATHWAYS

1.4.1   ACTIVATION OF MAP KINASES

MAP kinases mediate various immune responses, including activation of innate immunity and adaptive immunity.[68] TLR-mediated activation of MAP kinases can be analyzed by detecting the phosphorylation of serine/threonine or tyrosine residues of MAP kinases such as extracellular signal-regulated kinase (ERK; p38; or JNK. MAP kinases are activated through phosphorylation of threonine and tyrosine residues. On stimulation, the threonine and tyrosine residues of ERK, p38, and JNK at sequence T*EY* are phosphorylated by MAP kinase kinases. Phosphorylation of MAP kinases is monitored by Western blotting using antibodies that detect phosphorylated Thr202 and Tyr204 residues of ERK; phosphorylated Thr180 and Tyr182 residues of p38; and phosphorylated Thr183 and Tyr185 residues of JNK.[15]

Protocol 1.5: Monitoring MAP Kinase Phosphorylation by Western Blotting

1. Seed 10^6 cells into a 3.5 cm (for macrophages) or 6 cm (for DCs) dish.
2. Culture for over 1.5 hours so that cells become adherent to dish.
3. Stimulate cells with specific TLR ligands. Optimal concentration of TLR ligands: Pam3CSK4 (TLR1), 10 ng/ml; peptidoglycan (TLR2), 10 µg/ml; poly I:C (TLR3), 10 µg/ml; lipopolysaccharide (LPS; TLR4), 100 ng/ml; flagellin (TLR5), 10 µg/ml; mycoplasmal lipopeptide (FSL-1; TLR6), 10 ng/ml; Imiquimod (TLR7), 1 µg/ml; and CpG DNA (TLR9), 1 mM. All the TLR ligands are purchased from Invivogen (San Diego, CA).
4. Lyse cells in 200 µl of lysis buffer containing 1% TritonX-100, 20 mM Tris-HCl (pH 8.0), 137 mM NaCl, 5 mM EDTA, 10% glycerol, and protease inhibitor cocktail (Roche Diagnostics, Basel, Switzerland).
5. Add equal amounts of 2× sodium dodecyl sulfate polyacrylamide gel electrophoresis (SDS-PAGE) sample buffer containing 125 mM Tris-HCl (pH 6.5), 4% SDS, 20% glycerol, 0.02% bromophenol blue, and 4% 2-mercaptoethanol.
6. Apply 15 µl of sample per lane. Resolve proteins by SDS-PAGE on polyacrylamide gel (4–20% gradient gel), and transfer them onto a polyvinylidene difluoride (PVDF) membrane by electroblotting.

7. Blot the membrane with specific antibodies according to the manufacturer's instructions. Anti-JNK antibody, anti-ERK antibody, and anti-p38 antibodies are obtained from Santa Cruz Biotechnology (Santa Cruz, CA). Antibodies against phospho-JNK, phospho-ERK, and phospho-p38 are purchased from Cell Signaling Technology (Danvers, MA). These antibodies are used at 1:1,000 dilutions.

8. Visualize membrane-bound antibodies with horseradish peroxidase (HRP) conjugated antibody to rabbit IgG antibody (GE Healthcare UK, Buckinghamshire, UK) using the enhanced chemiluminescent system (PerkinElmer Life Sciences, Waltham, MA).

1.4.2 Activation of NF-κB

The transcription factor NF-κB is involved in the activation of various genes in response to infections, inflammation, and other stressful conditions. TLR stimulation of macrophages and DCs leads to the activation of NF-κB. NF-κB resides in the cytoplasm through association with its inhibitory molecule, IκB. Once stimulated, IκB is phosphorylated by the IKK complex, leading to proteasome-mediated degradation. NF-κB then becomes free and translocates into the nucleus, where it binds to specific sites of the promoter region of genes.[69] Thus, activity of NF-κB can be monitored by detecting the expression of IκB and the DNA binding of NF-κB.

Expression of IκB proteins, such as IκBα and JκBβ, is monitored by Western blotting using antibodies against IκBα and IκBβ, as described above. Before stimulation, expression of Iκ is detected. After stimulation, its expression is rapidly reduced through proteasome-dependent degradation of ubiquitinated IκB proteins. However, expression of IκB proteins is recovered promptly, since IκB itself is an NF-κ target gene and TLR-mediated NF-κB activation rapidly induces messenger RNA (mRNA) expression of the IκB gene. Thus, NF-κB activity can be analyzed by transient decrease in Iκ expression by Western blotting using antibodies against JκBα or JκBβ (Santa Cruz Biotechnology, Santa Cruz, CA).

Activity of NF-κB can be directly monitored by detecting its DNA binding activity in an electrophoretic mobility shift assay (EMSA).[15] EMSA is based on the observation that NF-κB–DNA complexes migrate more slowly than free DNA when subjected to nondenaturing PAGE. Because the rate of DNA migration is shifted or retarded on binding of NF-κB, the assay is also called a gel shift or gel retardation assay.

Protocol 1.6: Monitoring NF-κB Activity by Electrophoretic Mobility Shift Assay (EMSA)

1. Seed 1×10^6 cells in a 6 cm dish.
2. Stimulate cells with specific TLR ligands, and wash them with PBS.
3. Suspend cells in 400 μl of buffer A (10 mM HEPES [4-(2-hydroxyethyl)-1-piperazineethanesulfonic acid; pH 7.8], 10 mM KCl, 0.1 mM EDTA, 1

mM DTT [dithiothreitol], 2 µg/ml aprotinin, 0.5 mM phenylmethylsulfonyl fluoride [PMSF], and 0.5% Triton X-100).

4. Centrifuge samples, and pellet down nuclei in microcentrifuge tubes (5,000 rpm for 2 minutes at 4 °C).
5. Suspend the nuclei in buffer C (50 mM HEPES [pH 7.8], 420 mM KCl, 0.1 mM EDTA, 5 mM MgCl$_2$, 10% glycerol, 1 mM DTT, 2 µg/ml aprotinin, and 0.5 mM PMSF).
6. Vortex and stir the sample for 30 minutes at 4 °C.
7. Centrifuge the samples (15,000 rpm for 2 minutes at 4 °C), and transfer the supernatant containing nuclear proteins to a fresh vial.
8. Determine protein concentration of the nuclear extracts.
9. Label double-stranded, oligonucleotide probe–containing NF-κB-binding sites with [α-^{32}P]dCTP using an 3'-5'exonuclease-free Klenow fragment according to the manufacturer's instructions (Takara Bio, Otsu, Japan).
10. Incubate the nuclear extract (3 µg) with 300 femtomole (fmol) of the probe in a total of 25 µl of binding buffer (10 mM HEPES [pH 7.8], 50 mM KCl, 1 mM EDTA, 5 mM MgCl$_2$, 10% glycerol, and 2 µg of poly-dIdC) for 20 minutes at room temperature. Poly-dIdC is used to minimize the binding of nonspecific proteins to the labeled target DNA. These repetitive polymers provide an excess of nonspecific sites to adsorb proteins in crude lysates that will bind to any general DNA sequence.
11. Fractionate the samples on a 4% polyacrylamide gel in 25 mM Tris-HCl (pH 8.5), 190 mM glycine, and 1 mM EDTA.
12. Dry the gel, and visualize the signals by autoradiography.

To show that the retarded or shifted band stands for the specific binding to the target sequence, an unlabeled competitor oligonucleotide can be added to nuclear extracts 15 minutes before addition of the labeled probe. A fiftyfold molar excess of unlabeled target is usually sufficient to outcompete any specific interactions. While there are characteristic shifts caused by the binding of NF-κB to the target DNA, the change in relative mobility does not identify the bound NF-κB. The DNA-bound NF-κB can be visualized by addition of an antibody against NF-κB to the binding reaction. If NF-κB binds to the target DNA, the antibody will bind to that NF-κB–DNA complex, further decreasing its mobility relative to unbound DNA in what is called a "supershift." The supershift assay is performed by preincubation of nuclear extracts with 1 mg of anti-p65 antibody or anti-p50 antibody (Santa Cruz Biotechnology, Santa Cruz, CA) for 60 minutes at 4 °C before addition of the labeled probe.

1.4.3 Activation of IRF Transcription Factors

Transcription factors such as IRF-3 have potential phosphorylation sites in the C-terminal region (Ser385, 386 [2S site] and Ser396, 398, 402, 405, and Thr404 [5ST site] of human IRF-3). Phosphorylation of Ser396 of IRF-3 upon TLR3 or TLR4 stimulation was demonstrated by using phosphospecific antibody.[70] Other reports have shown that phosphorylation of Ser386 is also a critical determinant of the activation of IRF-3.[71,72] No direct evidence of phosphorylation has been reported for the remaining five serine/threonine sites. Phosphorylation of IRF-3 induces its activation, homodimerization, and nuclear translocation. Therefore, activation of IRF-3 can be monitored by detection of nuclear translocation or homodimerization of IRF-3.[73]

Protocol 1.7: Monitoring IRF-3 Nuclear Translocation by Western Blotting

1. Stimulate cells (3×10^6) with specific TLR ligands, wash them with PBS, and suspend them in 1 ml of PBS.
2. Pellet down the cells, and resuspend them in 150 μl of buffer A (10 mM HEPES-KOH [pH 7.8], 10mM KCl, 0.1 mM EDTA [pH 8.0], 1 μg/ml aprotonin, 1 μg/ml pepstatin, and 0.5 mM PMSF).
3. Add NP-40 (to final concentration 0.2%) and incubate at room temperature for 5 minutes. This allows disruption of plasma membrane and release of nuclei.
4. Pellet down the nuclei, and wash them twice with 800 μl of buffer A (without NP-40).
5. Resuspend the nuclei in 40–80 μl of buffer C (50 mM HEPES-KOH [pH 7.8], 420 mM KCl, 5 mM MgCl$_2$, 0.1 mM EDTA [pH 8.0], and 2% glycerol) and incubate them at 40 °C for 30 minutes to allow nuclear protein extraction.
6. Determine protein concentration of the nuclear extracts, and use 30–50 μg of proteins for Western blotting as described above. Anti-IRF-3 antibody can be purchased from BD Biosciences (San Jose, CA).

Protocol 1.8: Monitoring Homodimerization of IRF-3

In SDS-PAGE, the conformation or complex structure of samples is destroyed by the addition of SDS and 2-mercaptoethanol. In contrast, the native-PAGE technique allows samples to be electrophoresed in polyacrylamide gel, keeping their natural conformation. The mobility of IRF-3 homodimers is retarded compared with that of nonactivated single monomers of IRF-3, thereby indicating the active state.

1. Stimulate cells (3×10^6) with specific TLR ligands, and wash them with PBS.
2. Lyse cells in 100 μl of lysis buffer (50 mM Tris-HCl [pH 8.0], 1% Nonidet P-40, 150 mM NaCl, 100 μg/ml leupeptin, 1 mM PMSF, and 5 mM orthovanadate).
3. Centrifuge the lysate at 10,000 rpm for 5 minutes to remove the insoluble fraction.

4. Mix the sample (2 µl, 10 µg of protein) with 2 µl of 2× loading buffer (125 mM Tris-HCl [pH 6.8], 30% glycerol, and 0.002% bromphenol blue).

5. Prerun 7.5% polyacrylamide gel in running buffer containing 25 mM Tris-HCl (pH 8.4) and 192 mM glycine with and without 1% deoxycholate in the cathode and anode chamber, respectively, for 30 minutes at 40 mA.

6. Apply the sample to the gel, and electrophorese for 60 minutes at 25 mA at 4 °C.

7. Transfer the proteins onto a PVDF membrane by electroblotting.

8. Blot the PVDF membrane with anti-IRF-3 antibody (BD Biosciences, San Jose, CA).

9. Visualize the membrane-bound antibodies with HRP-conjugated antibody to rabbit IgG (GE Healthcare UK, Buckinghamshire, UK) using the enhanced chemiluminescent system (PerkinElmer Life Sciences, Waltham, MA).

Other members of the IRF family of transcription factors, such as IRF-5 and IRF-7, show similar patterns of activation.

REFERENCES

1. Lemaitre, B., Nicolas, E., Michaut, L., Reichhart, J. M., and Hoffmann, J. A. The dorsoventral regulatory gene cassette spatzle/Toll/cactus controls the potent antifungal response in Drosophila adults. Cell 86, 973, 1996.

2. Medzhitov, R., Preston-Hurlburt, P., and Janeway, C. A., Jr. A human homologue of the Drosophila Toll protein signals activation of adaptive immunity. Nature 388, 394, 1997.

3. Takeda, K., Kaisho, T., and Akira, S. Toll-like receptors. Annu Rev Immunol 21, 335, 2003.

4. Iwasaki, A., and Medzhitov, R. Toll-like receptor control of the adaptive immune responses. Nat Immunol 5, 987, 2004.

5. Heil, F., Ahmad-Nejad, P., Hemmi, H., Hochrein, H., Ampenberger, F., Gellert, T., Dietrich, H., Lipford, G., Takeda, K., Akira, S., et al. The Toll-like receptor 7 (TLR7)-specific stimulus loxoribine uncovers a strong relationship within the TLR7, 8 and 9 subfamily. Eur J Immunol 33, 2987, 2003.

6. Matsumoto, M., Funami, K., Tanabe, M., Oshiumi, H., Shingai, M., Seto, Y., Yamamoto, A., and Seya, T. Subcellular localization of Toll-like receptor 3 in human dendritic cells. J Immunol 171, 3154, 2003.

7. Ahmad-Nejad, P., Hacker, H., Rutz, M., Bauer, S., Vabulas, R. M., and Wagner, H. Bacterial CpG-DNA and lipopolysaccharides activate Toll-like receptors at distinct cellular compartments. Eur J Immunol 32, 1958, 2002.

8. Latz, E., Schoenemeyer, A., Visintin, A., Fitzgerald, K. A., Monks, B. G., Knetter, C. F., Lien, E., Nilsen, N. J., Espevik, T., and Golenbock, D. T. TLR9 signals after translocating from the ER to CpG DNA in the lysosome. Nat Immunol 5, 190, 2004.

9. Heil, F., Hemmi, H., Hochrein, H., Ampenberger, F., Kirschning, C., Akira, S., Lipford, G., Wagner, H., and Bauer, S. Species-specific recognition of single-stranded RNA via Toll-like receptor 7 and 8. Science 303, 1526, 2004.

10. Diebold, S. S., Kaisho, T., Hemmi, H., Akira, S., and Reis e Sousa, C. Innate antiviral responses by means of TLR7-mediated recognition of single-stranded RNA. Science 303, 1529, 2004.

11. Lund, J. M., Alexopoulou, L., Sato, A., Karow, M., Adams, N. C., Gale, N. W., Iwasaki, A., and Flavell, R. A. Recognition of single-stranded RNA viruses by Toll-like receptor 7. Proc Natl Acad Sci USA 101, 5598, 2004.

12. Hacker, H., Mischak, H., Miethke, T., Liptay, S., Schmid, R., Sparwasser, T., Heeg, K., Lipford, G. B., and Wagner, H. CpG-DNA-specific activation of antigen-presenting cells requires stress kinase activity and is preceded by non-specific endocytosis and endosomal maturation. Embo J 17, 6230, 1998.

13. Underhill, D. M., Ozinsky, A., Hajjar, A. M., Stevens, A., Wilson, C. B., Bassetti, M., and Aderem, A. The Toll-like receptor 2 is recruited to macrophage phagosomes and discriminates between pathogens. Nature 401, 811, 1999.

14. Saitoh, S., Akashi, S., Yamada, T., Tanimura, N., Kobayashi, M., Konno, K., Matsumoto, F., Fukase, K., Kusumoto, S., Nagai, Y., et al. Lipid A antagonist, lipid IVa, is distinct from lipid A in interaction with Toll-like receptor 4 (TLR4)-MD-2 and ligand-induced TLR4 oligomerization. Int Immunol 16, 961, 2004.

15. Kawai, T., Adachi, O., Ogawa, T., Takeda, K., and Akira, S. Unresponsiveness of MyD88-deficient mice to endotoxin. Immunity 11, 115, 1999.

16. Takeuchi, O., Takeda, K., Hoshino, K., Adachi, O., Ogawa, T., and Akira, S. Cellular responses to bacterial cell wall components are mediated through MyD88-dependent signaling cascades. Int Immunol 12, 113, 2000.

17. Toshchakov, V., Jones, B. W., Perera, P. Y., Thomas, K., Cody, M. J., Zhang, S., Williams, B. R., Major, J., Hamilton, T. A., Fenton, M. J., and Vogel, S. N. TLR4, but not TLR2, mediates IFN-beta-induced STAT1alpha/beta-dependent gene expression in macrophages. Nat Immunol 3, 392, 2002.

18. Doyle, S., Vaidya, S., O'Connell, R., Dadgostar, H., Dempsey, P., Wu, T., Rao, G., Sun, R., Haberland, M., Modlin, R., and Cheng, G. IRF3 mediates a TLR3/TLR4-specific antiviral gene program. Immunity 17, 251, 2002.

19. Hemmi, H., Kaisho, T., Takeda, K., and Akira, S. The roles of Toll-like receptor 9, MyD88, and DNA-dependent protein kinase catalytic subunit in the effects of two distinct CpG DNAs on dendritic cell subsets. J Immunol 170, 3059, 2003.

20. Ito, T., Amakawa, R., Kaisho, T., Hemmi, H., Tajima, K., Uehira, K., Ozaki, Y., Tomizawa, H., Akira, S., and Fukuhara, S. Interferon-alpha and interleukin-12 are induced differentially by Toll-like receptor 7 ligands in human blood dendritic cell subsets. J Exp Med 195, 1507, 2002.

21. Honda, K., Takaoka, A., and Taniguchi, T. Type I interferon gene induction by the interferon regulatory factor family of transcription factors. Immunity 25, 349, 2006.

22. Sato, S., Sanjo, H., Takeda, K., Ninomiya-Tsuji, J., Yamamoto, M., Kawai, T., Matsumoto, K., Takeuchi, O., and Akira, S. Essential function for the kinase TAK1 in innate and adaptive immune responses. Nat Immunol 6, 1087, 2005.

23. Yamamoto, M., Okamoto, T., Takeda, K., Sato, S., Sanjo, H., Uematsu, S., Saitoh, T., Yamamoto, N., Sakurai, H., Ishii, K. J., et al. Key function for the Ubc13 E2 ubiquitin-conjugating enzyme in immune receptor signaling. Nat Immunol 7, 962, 2006.

24. Takaoka, A., Yanai, H., Kondo, S., Duncan, G., Negishi, H., Mizutani, T., Kano, S., Honda, K., Ohba, Y., Mak, T. W., and Taniguchi, T. Integral role of IRF-5 in the gene induction programme activated by Toll-like receptors. Nature 434, 243, 2005.

25. Honda, K., Yanai, H., Mizutani, T., Negishi, H., Shimada, N., Suzuki, N., Ohba, Y., Takaoka, A., Yeh, W. C., and Taniguchi, T. Role of a transductional-transcriptional processor complex involving MyD88 and IRF-7 in Toll-like receptor signaling. Proc Natl Acad Sci USA 101, 15416, 2004.

26. Kawai, T., Sato, S., Ishii, K. J., Coban, C., Hemmi, H., Yamamoto, M., Terai, K., Matsuda, M., Inoue, J., Uematsu, S., et al. Interferon-alpha induction through Toll-like receptors involves a direct interaction of IRF7 with MyD88 and TRAF6. Nat Immunol 5, 1061, 2004.

27. Honda, K., Yanai, H., Negishi, H., Asagiri, M., Sato, M., Mizutani, T., Shimada, N., Ohba, Y., Takaoka, A., Yoshida, N., and Taniguchi, T. IRF-7 is the master regulator of type-I interferon-dependent immune responses. Nature 434, 772, 2005.

28. Negishi, H., Ohba, Y., Yanai, H., Takaoka, A., Honma, K., Yui, K., Matsuyama, T., Taniguchi, T., and Honda, K. Negative regulation of Toll-like-receptor signaling by IRF-4. Proc Natl Acad Sci USA 102, 15989, 2005.

29. Negishi, H., Fujita, Y., Yanai, H., Sakaguchi, S., Ouyang, X., Shinohara, M., Takayanagi, H., Ohba, Y., Taniguchi, T., and Honda, K. Evidence for licensing of IFN-gamma-induced IFN regulatory factor 1 transcription factor by MyD88 in Toll-like receptor-dependent gene induction program. Proc Natl Acad Sci USA 103, 15136, 2006.

30. Zhao, J., Kong, H. J., Li, H., Huang, B., Yang, M., Zhu, C., Bogunovic, M., Zheng, F., Mayer, L., Ozato, K., et al. IRF-8/interferon (IFN) consensus sequence-binding protein is involved in Toll-like receptor (TLR) signaling and contributes to the cross-talk between TLR and IFN-gamma signaling pathways. J Biol Chem 281, 10073, 2006.

31. Horng, T., Barton, G. M., and Medzhitov, R. TIRAP: an adapter molecule in the Toll signaling pathway. Nat Immunol 2, 835, 2001.

32. Fitzgerald, K. A., Palsson-McDermott, E. M., Bowie, A. G., Jefferies, C. A., Mansell, A. S., Brady, G., Brint, E., Dunne, A., Gray, P., Harte, M. T., et al. Mal (MyD88-adapter-like) is required for Toll-like receptor-4 signal transduction. Nature 413, 78, 2001.

33. Horng, T., Barton, G. M., Flavell, R. A., and Medzhitov, R. The adaptor molecule TIRAP provides signalling specificity for Toll-like receptors. Nature 420, 329, 2002.

34. Yamamoto, M., Sato, S., Hemmi, H., Sanjo, H., Uematsu, S., Kaisho, T., Hoshino, K., Takeuchi, O., Kobayashi, M., Fujita, T., et al. Essential role for TIRAP in activation of the signalling cascade shared by TLR2 and TLR4. Nature 420, 324, 2002.

35. Colonna, M., Trinchieri, G., and Liu, Y. J. Plasmacytoid dendritic cells in immunity. Nat Immunol 5, 1219, 2004.

36. Akira, S., Uematsu, S., and Takeuchi, O. Pathogen recognition and innate immunity. Cell 124, 783, 2006

37. Uematsu, S., Sato, S., Yamamoto, M., Hirotani, T., Kato, H., Takeshita, F., Matsuda, M., Coban, C., Ishii, K. J., Kawai, T., et al. Interleukin-1 receptor-associated kinase-1 plays an essential role for Toll-like receptor (TLR)7- and TLR9-mediated interferon-{alpha} induction. J Exp Med 201, 915, 2005.

38. Hoshino, K., Sugiyama, T., Matsumoto, M., Tanaka, T., Saito, M., Hemmi, H., Ohara, O., Akira, S., and Kaisho, T. IkappaB kinase-alpha is critical for interferon-alpha production induced by Toll-like receptors 7 and 9. Nature 440, 949, 2006.

39. Honda, K., Ohba, Y., Yanai, H., Negishi, H., Mizutani, T., Takaoka, A., Taya, C., and Taniguchi, T. Spatiotemporal regulation of MyD88-IRF-7 signalling for robust type-I interferon induction. Nature 434, 1035, 2005.

40. Kawai, T., Takeuchi, O., Fujita, T., Inoue, J., Muhlradt, P. F., Sato, S., Hoshino, K., and Akira, S. Lipopolysaccharide stimulates the MyD88-independent pathway and results in activation of IFN-regulatory factor 3 and the expression of a subset of lipopolysaccharide-inducible genes. J Immunol 167, 5887, 2001.

41. Yamamoto, M., Sato, S., Mori, K., Hoshino, K., Takeuchi, O., Takeda, K., and Akira, S. Cutting edge: a novel Toll/IL-1 receptor domain-containing adapter that preferentially activates the IFN-beta promoter in the Toll-like receptor signaling. J Immunol 169, 6668, 2002.

42. Oshiumi, H., Matsumoto, M., Funami, K., Akazawa, T., and Seya, T. TICAM-1, an adaptor molecule that participates in Toll-like receptor 3-mediated interferon-beta induction. Nat Immunol 4, 161, 2003.

43. Yamamoto, M., Sato, S., Hemmi, H., Hoshino, K., Kaisho, T., Sanjo, H., Takeuchi, O., Sugiyama, M., Okabe, M., Takeda, K., and Akira, S. Role of adaptor TRIF in the MyD88-independent Toll-like receptor signaling pathway. Science 301, 640, 2003.
44. Hoebe, K., Du, X., Georgel, P., Janssen, E., Tabeta, K., Kim, S. O., Goode, J., Lin, P., Mann, N., Mudd, S., et al. Identification of Lps2 as a key transducer of MyD88-independent TIR signalling. Nature 424, 743, 2003.
45. Fitzgerald, K. A., Rowe, D. C., Barnes, B. J., Caffrey, D. R., Visintin, A., Latz, E., Monks, B., Pitha, P. M., and Golenbock, D. T. LPS-TLR4 signaling to IRF-3/7 and NF-kappaB involves the Toll adapters TRAM and TRIF. J Exp Med 198, 1043, 2003.
46. Yamamoto, M., Sato, S., Hemmi, H., Uematsu, S., Hoshino, K., Kaisho, T., Takeuchi, O., Takeda, K., and Akira, S. TRAM is specifically involved in the Toll-like receptor 4-mediated MyD88-independent signaling pathway. Nat Immunol 4, 1144, 2003.
47. Sharma, S., tenOever, B. R., Grandvaux, N., Zhou, G. P., Lin, R., and Hiscott, J. Triggering the interferon antiviral response through an IKK-related pathway. Science 300, 1148, 2003.
48. Fitzgerald, K. A., McWhirter, S. M., Faia, K. L., Rowe, D. C., Latz, E., Golenbock, D. T., Coyle, A. J., Liao, S. M., and Maniatis, T. IKKepsilon and TBK1 are essential components of the IRF3 signaling pathway. Nat Immunol 4, 491, 2003.
49. McWhirter, S. M., Fitzgerald, K. A., Rosains, J., Rowe, D. C., Golenbock, D. T., and Maniatis, T. IFN-regulatory factor 3-dependent gene expression is defective in Tbk1-deficient mouse embryonic fibroblasts. Proc Natl Acad Sci USA 101, 233, 2004.
50. Hemmi, H., Takeuchi, O., Sato, S., Yamamoto, M., Kaisho, T., Sanjo, H., Kawai, T., Hoshino, K., Takeda, K., and Akira, S. The roles of two IkappaB kinase-related kinases in lipopolysaccharide and double stranded RNA signaling and viral infection. J Exp Med 199, 1641, 2004.
51. Perry, A. K., Chow, E. K., Goodnough, J. B., Yeh, W. C., and Cheng, G. Differential requirement for TANK-binding kinase-1 in type I interferon responses to Toll-like receptor activation and viral infection. J Exp Med 199, 1651, 2004.
52. Meylan, E., Burns, K., Hofmann, K., Blancheteau, V., Martinon, F., Kelliher, M., and Tschopp, J. RIP1 is an essential mediator of Toll-like receptor 3-induced NF-kappa B activation. Nat Immunol 5, 503, 2004.
53. Hacker, H., Redecke, V., Blagoev, B., Kratchmarova, I., Hsu, L. C., Wang, G. G., Kamps, M. P., Raz, E., Wagner, H., Hacker, G., et al. Specificity in Toll-like receptor signalling through distinct effector functions of TRAF3 and TRAF6. Nature 439, 204, 2006.
54. Oganesyan, G., Saha, S. K., Guo, B., He, J. Q., Shahangian, A., Zarnegar, B., Perry, A., and Cheng, G. Critical role of TRAF3 in the Toll-like receptor-dependent and -independent antiviral response. Nature 439, 208, 2006.
55. Gordon, S., and Taylor, P. R. Monocyte and macrophage heterogeneity. Nat Rev Immunol 5, 953, 2005.
56. Banchereau, J., Briere, F., Caux, C., Davoust, J., Lebecque, S., Liu, Y. J., Pulendran, B., and Palucka, K. Immunobiology of dendritic cells. Annu Rev Immunol 18, 767, 2000.
57. Heath, W. R., Belz, G. T., Behrens, G. M., Smith, C. M., Forehan, S. P., Parish, I. A., Davey, G. M., Wilson, N. S., Carbone, F. R., and Villadangos, J. A. Cross-presentation, dendritic cell subsets, and the generation of immunity to cellular antigens. Immunol Rev 199, 9, 2004.
58. Grouard, G., Rissoan, M. C., Filgueira, L., Durand, I., Banchereau, J., and Liu, Y. J. The enigmatic plasmacytoid T cells develop into dendritic cells with interleukin (IL)-3 and CD40-ligand. J Exp Med 185, 1101, 1997.

59. O'Keeffe, M., Hochrein, H., Vremec, D., Caminschi, I., Miller, J. L., Anders, E. M., Wu, L., Lahoud, M. H., Henri, S., Scott, B., et al. Mouse plasmacytoid cells: long-lived cells, heterogeneous in surface phenotype and function, that differentiate into CD8(+) dendritic cells only after microbial stimulus. J Exp Med 196, 1307, 2002.

60. Asselin-Paturel, C., Boonstra, A., Dalod, M., Durand, I., Yessaad, N., Dezutter-Dambuyant, C., Vicari, A., O'Garra, A., Biron, C., Briere, F., and Trinchieri, G. Mouse type I IFN-producing cells are immature APCs with plasmacytoid morphology. Nat Immunol 2, 1144, 2001.

61. Nakano, H., Yanagita, M., and Gunn, M. D. CD11c(+)B220(+)Gr-1(+) cells in mouse lymph nodes and spleen display characteristics of plasmacytoid dendritic cells. J Exp Med 194, 1171, 2001.

62. Le Bon, A., Etchart, N., Rossmann, C., Ashton, M., Hou, S., Gewert, D., Borrow, P., and Tough, D. F. Cross-priming of CD8+ T cells stimulated by virus-induced type I interferon. Nat Immunol 4, 1009, 2003.

63. Henri, S., Vremec, D., Kamath, A., Waithman, J., Williams, S., Benoist, C., Burnham, K., Sacland, S., Handman, E., and Shortman, K. The dendritic cell populations of mouse lymph nodes. J Immunol 167, 741, 2001.

64. Inaba, K., Inaba, M., Romani, N., Aya, H., Deguchi, M., Ikehara, S., Muramatsu, S., and Steinman, R. M. Generation of large numbers of dendritic cells from mouse bone marrow cultures supplemented with granulocyte/macrophage colony-stimulating factor. J Exp Med 176, 1693, 1992.

65. Pulendran, B., Banchereau, J., Burkeholder, S., Kraus, E., Guinet, E., Chalouni, C., Caron, D., Maliszewski, C., Davoust, J., Fay, J., and Palucka, K. Flt3-ligand and granulocyte colony-stimulating factor mobilize distinct human dendritic cell subsets in vivo. J Immunol 165, 566, 2000.

66. Gilliet, M., Boonstra, A., Paturel, C., Antonenko, S., Xu, X. L., Trinchieri, G., O'Garra, A., and Liu, Y. J. The development of murine plasmacytoid dendritic cell precursors is differentially regulated by FLT3-ligand and granulocyte/macrophage colony-stimulating factor. J Exp Med 195, 953, 2002.

67. Brawand, P., Fitzpatrick, D. R., Greenfield, B. W., Brasel, K., Maliszewski, C. R., and De Smedt, T. Murine plasmacytoid pre-dendritic cells generated from Flt3 ligand-supplemented bone marrow cultures are immature APCs. J Immunol 169, 6711, 2002.

68. Dong, C., Davis, R. J., and Flavell, R. A. MAP kinases in the immune response. Annu Rev Immunol 20, 55, 2002.

69. Ghosh, S., May, M. J., and Kopp, E. B. NF-kappa B and Rel proteins: evolutionarily conserved mediators of immune responses. Annu Rev Immunol 16, 225, 1998.

70. Servant, M. J., Grandvaux, N., tenOever, B. R., Duguay, D., Lin, R., and Hiscott, J. Identification of the minimal phosphoacceptor site required for in vivo activation of interferon regulatory factor 3 in response to virus and double-stranded RNA. J Biol Chem 278, 9441, 2003.

71. Mori, M., Yoneyama, M., Ito, T., Takahashi, K., Inagaki, F., and Fujita, T. Identification of Ser-386 of interferon regulatory factor 3 as critical target for inducible phosphorylation that determines activation. J Biol Chem 279, 9698, 2004.

72. Saitoh, T., Yamamoto, M., Miyagishi, M., Taira, K., Nakanishi, M., Fujita, T., Akira, S., Yamamoto, N., and Yamaoka, S. A20 is a negative regulator of IFN regulatory factor 3 signaling. J Immunol 174, 1507, 2005.

73. Iwamura, T., Yoneyama, M., Yamaguchi, K., Suhara, W., Mori, W., Shiota, K., Okabe, Y., Namiki, H., and Fujita, T. Induction of IRF-3/-7 kinase and NF-kappaB in response to double-stranded RNA and virus infection: common and unique pathways. Genes Cells 6, 375, 2001.

2 TLR Ligands in Experimental Settings
Their Purity and Specificity

Min-Fu Tsan

CONTENTS

2.1 INTRODUCTION

Since the identification of mammalian Toll-like receptors (TLRs) by Medzhitov et al.[1] in 1997, TLRs have been shown to play a crucial role in the innate host defense against invading microorganisms by recognizing conserved motifs of microbial origin, also known as pathogen-associated molecular patterns (PAMPs).[2,3] Recent reports of endogenous molecules of mammalian origin such as the endogenous ligands of TLRs further expand the potential role of TLRs into the pathogenesis of autoimmune diseases and chronic sterile inflammatory disorders.[4–6] However, exactly how 11 members of mammalian TLRs recognize such a diversity of molecular structures is not clear.

2.1.1 TOLL-LIKE RECEPTORS

Toll-like receptors, the mammalian homologues of the *Drosophila* Toll protein, are members of the interleukin-1 receptor (IL-1R) superfamily that share significant homology with a Toll/IL-1R (TIR) domain in the cytoplasmic region and a leucine-rich repeat (LRR) domain in the extracellular region.[2,7] The cytoplasmic TIR domain is responsible for intracellular signaling via recruitment of adaptor molecules, such as myeloid differentiation factor 88 (MyD88), TIR-associated protein (TIRAP), TIR domain–containing adaptor inducing interferon-β (TRIF), and TRIF-related adaptor molecule (TRAM), leading to the activation of transcription factors, nuclear factor κB (NF-κB) and interferon (IFN)-regulated factors (IRFs), and to the subsequent production of proinflammatory cytokines and type I IFNs.[7,8] The extracellular LRR domain, on the other hand, may play a critical role in the recognition of TLR ligands. For example, the crystallographic structure of TLR3 reveals a large horseshoe-shaped solenoid composed of 23 LRRs.[9,10] Analysis of the unliganded TLR3 structure reveals two patches of basic residues and two binding sites for phosphate backbone mimic, sulfate ions, that may be capable of recognizing the TLR3 ligand, double-stranded RNA (dsRNA).[11] On the other hand, it has been shown that lipopolysaccharide (LPS), the TLR4 ligand, binds to the accessory molecule, myeloid differentiation protein-2 (MD-2), in the TLR4/MD-2 receptor complex, but does not bind TLR4 directly.[12,13] Thus, further crystallographic studies of liganded TLRs will be necessary to elucidate the role of LRR domain in TLR ligand recognition.

2.1.2 TLR LIGANDS

The identification of TLR ligands is usually carried out using TLR-deficient cells derived from *tlr* knockout mice or natural mutants, and/or cells transfected with TLR complementary DNAs (cDNAs).[2] Anti-TLR antibodies can also be used if the TLR is present on the cell surface. However, inhibition of TLR by anti-TLR antibodies may not be complete.

The subcellular localization of TLRs also plays an important role in the TLR ligand recognition. For example, TLR1, TLR2, TLR4, and TLR6 are located on the cell surface. They are capable of interacting with their respective ligands at the cell surface.[2] On the other hand, TLR3, TLR7, TLR8, and TLR9 are not expressed on the cell surface in most cells. They interact with their respective ligands in endosomal or lysosomal compartments.[3,14,15] Unlike single-stranded RNAs (ssRNAs), the natural ligand of TLR7 and TLR8, small interfering RNAs (siRNAs) can activate TLR7 and TLR8 only when they are encapsulated in liposomes.[16] Likewise, mRNAs can activate TLR3 only when they are complexed with lipofectin.[17] Thus, the ability of their ligands to gain access to endosomal or lysosomal compartments is an essential first step in the activation of TLR3, TLR7, TLR8, and TLR9.[3,18]

Based on the types of PAMPs they recognize, TLRs can be grouped into at least three families.[7,8] First, TLR2 (in combination with TLR1 or TLR6 as heterodimers) and TLR4 (as a homodimer) recognize lipid-based ligands, such as bacterial lipoproteins or lipoteichoic acid (TLR2)[19,20] and LPS (TLR4).[21] Second, TLR3, TLR7, TLR8, and TLR9 recognize bacterial and viral nucleic acids, such as dsRNA (TLR3),[22] ssRNA (TLR7 and TLR8),[23,24] and nonmethylated CpG DNA (TLR9).[25]

Third, TLR5 and TLR11 recognize microbial proteins such as flagellin (TLR5)[26] and profilin (murine TLR11).[27] No ligand has been identified for TLR10. Obviously, this classification is an oversimplification. As shown in Table 2.1, TLRs, particularly TLR2 and TLR4, have been reported to recognize divergent molecules of microbial and mammalian origins. In addition to LPS, TLR4 has been reported to recognize various microbial and mammalian proteins (e.g., respiratory syncythia virus envelope fusion protein,[28] pancreatic elastase,[29] and lipoprotein [i.e., minimally modified or oxidized low-density lipoprotein, or mmLDL])[30] and polysaccharides (e.g., heparan sulfate and soluble hyaluronan).[31,32] Furthermore, some endogenous molecules are reported to be ligands for both TLR2 and TLR4.[5,33,34] The reason for this promiscuity of TLRs is not clear. However, theories abound to explain these observations.

The danger theory[35] proposes that the immune system has evolved primarily to recognize danger signals, both exogenous and endogenous, rather than just the non-self signals as originally proposed in the self-nonself immune surveillance hypothesis.[36] The surveillance model proposes that the immune system recognizes not only the exogenous and endogenous molecules but also the degradation products of endogenous macromolecules, which indicate tissue injury, infection, and/or tissue remodeling.[37] The more recent hydrophobicity model proposes that immune receptors such as TLRs are evolved to recognize the hydrophobic portions of endogenous as well as exogenous biological molecules.[38] The hydrophobic portions of these endogenous molecules are normally protected from being exposed to the immune system. However, in situations where intracellular molecules are released due to cell necrosis or trauma, or the released intracellular molecules and extracellular molecules are modified or degraded, the hydrophobic portions of these molecules become accessible to the immune system to initiate repair, remodeling, or immune response. Thus, according to this model, any molecules with hydrophobic portions, when exposed, can become damage-associated molecular patterns (DAMPs) and are recognized by TLRs. These DAMPs encompass PAMPs of microbial origin as well as danger signals of endogenous origin.[38] A new term, "alarmin," has also been coined for danger signals of endogenous origin.[39]

2.1.3 PAMP CONTAMINATION AS PUTATIVE TLR LIGANDS

As bacterial cell wall products, particularly LPS, are ubiquitous and are frequent contaminants of purified commercial and noncommercial preparations, another possibility to account for the reported promiscuity of TLR2 and TLR4 is the contamination of putative TLR2 and TLR4 ligands by PAMPs. In fact, this has been shown to be the case in a number of putative TLR2 and TLR4 ligands.[5,34]

2.1.3.1 Heat Shock Proteins

Mammalian heat shock proteins (HSPs) such as Hsp60, Hsp70, and gp96 (glucose-regulated protein 96, the endoplasmic reticulum form of Hsp90) have been reported to induce proinflammatory cytokine production via TLR2 and TLR4.[40–43] However, when HSP preparations that have been purified to be essentially free of LPS contamination are used, no HSP cytokine effects can be demonstrated. This is the case for Hsp60, Hsp70, Hsp90, and gp96.[43–52] The fact that Hsp60, Hsp70, and gp96 were

TABLE 2.1

Reported Toll-Like Receptor Ligands[a]

Receptor	Exogenous Ligands	Synthetic Ligands	Endogenous Ligands
TLR 1	Bacterial lipoproteins	Triacyl lipopeptides	
TLR2	Lipoproteins and lipopeptides Peptidoglycan Lipoteichoic acid Lipoarabinomannan Glycosylphosphatidyl inositol Zymosan Leptospiral lipopolysaccharide	Diacyl and triacyl lipopeptides	
TLR3	Double-stranded DNA (dsRNA)	Poly I:C	
TLR4	Lipopolysaccharide (LPS) Taxol Respiratory syncythia virus F protein Tamm-Horsfall glycoprotein	Synthetic lipid A	Fibrinogen Surfactant protein-A[b] Fibrinogen extra domain A Heparan sulfate Soluble hyaduronan β-defensin 2-fusion protein Minimally modified (or oxidized) low-density lipoprotein (mmLDL) Pancreatic elastase[b] Hsp22 αA crystalline
TLR2 and TLR4	Hsp60 and Hsp70		Hsp60,[b] Hsp70,[b] and gp96[b] High-mobility group box-1 protein (HMGB-1)[b] Urate crystal

TLR5	Bacterial flagellin		
TLR6	Mycobacterial lipopeptides Zymosan	Diacyl lipopeptides	
TLR7	Single-stranded RNA (ssRNA)	Imidazoquinolines	RNA immune complex
TLR8	ssRNA	Imidazoquinolines	
TLR9	Bacterial DNA Hemozoin (malaria pigment)	CpG oligonucleotide	Chromatin immune complex DNA immune complex
TLR10	Not determined		
TLR11	Profilin		

a. Data are summarized from References 2, 3, 5–8, and 34.

b. These reported TLR ligands have been shown to be due to PAMP contamination.

reported to be endogenous ligands of TLR2 and TLR4 suggested that these HSP preparations were contaminated not only with LPS but also with TLR2 ligand(s) of microbial origin. Gao and Tsan[47] demonstrated that 50% of the TNFα-inducing activity of the commercially available rhHsp60 (#NSP-540, StressGene Biotechnologies, Victoria, BC, Canada) was due to a non-LPS contaminant(s) that was heat sensitive but not inhibitable by polymyxin B. The exact nature of this non-LPS contaminant(s) is not clear. However, it was designated as an LPS-associated molecule(s), since it could be removed along with LPS using polymyxin B agarose column. Recently, Ye and Gan[53] reported that in the presence of an anti-CD₃ antibody, *Burkholderia pseudomallei* Hsp70 and *Mycobactrium tuberculosis* Hsp70 were able to stimulate Jurkat T cells to produce IL-12 via TLR5. However, the co-stimulatory effect of these bacterial Hsp70s was due to flagellin contamination.

2.1.3.2   Pancreatic Elastase

It has been suggested that pancreatic elastase may be responsible for the systemic inflammatory response syndrome (SIRS) associated with severe acute pancreatitis, because commercially available pancreatic elastase is capable of inducing the activation of macrophages in vitro and SIRS in vivo in mice.[54] Furthermore, the cytokine effects of pancreatic elastase are shown to be mediated by TLR4.[28] However, Geisler et al.[55] recently demonstrated that highly purified, low-endotoxin pancreatic elastase preparation, which retained its full elastase enzyme activity, failed to show any cytokine effects in vitro and in vivo. In contrast, the less purified commercially available pancreatic elastase preparation that was shown to induce macrophage activation in vitro and SIRS in vivo contained LPS at a concentration that was three orders of magnitude higher than that of the highly purified low-endotoxin preparation. These results suggest that pancreatic elastase is not a TLR4 ligand and is not responsible for the SIRS in acute pancreatitis.

2.1.3.3   C-Reactive Protein

C-reactive protein is an acute-phase protein. Recently, there has been considerable interest in the role of C-reactive protein in the pathogenesis of atherosclerosis.[56] It has been shown that C-reactive protein is capable of inducing the activation of endothelial cells and macrophages in vitro.[57,58] In addition, intravenous administration of commercially available recombinant human C-reactive protein into human subjects leads to systemic inflammation and activation of the coagulation system.[59] Thus, it has been suggested that C-reactive protein is not just an innocent bystander but also a pivotal mediator of atherosclerosis and atherothrombotic events.[56] However, Pepys et al.[60] recently demonstrated that in contrast to the commercially available recombinant human C-reactive protein, highly purified native human C-reactive protein did not have the capacity to induce macrophage activation in vitro or in vivo in mice. The commercial recombinant human C-reactive protein preparation that was shown to have proinflammatory and procoagulation activities was found to contain a high level of LPS contamination. Thus, C-reactive protein is unlikely to play a significant role in the pathogenesis of atherosclerosis.

2.1.3.4 Surfactant Protein-A

Surfactant protein-A (SP-A), the most abundant surfactant protein present in the alveolar space, has been reported to be the endogenous ligand of TLR4.[61] However, the preparation of SP-A studied contained 140 pg LPS per μg of the surfactant protein. With the concentrations of 2.5–20 μg/ml SP-A tested, the final concentrations of LPS (350–2,800 pg/ml) are sufficient to account for the observed cytokine effects.[5] Recently, Yamada et al.[62] were unable to demonstrate the activation of TLR4 by SP-A. As a matter of fact, they showed that SP-A bound to TLR4 and MD-2, and inhibited smooth LPS-induced activation of NF-κB and production of TNFα. Thus, the reported cytokine effects of SP-A is likely due to LPS contamination.

2.1.3.5 High-Mobility Group Box-1 Protein

High-mobility group box-1 (HMGB1) protein is a nonhistone, DNA-binding nuclear protein that plays a critical role in regulating gene transcription.[63] HMGB1 protein has also been shown to act as a late mediator of endotoxic shock and to exert a diversity of proinflammatory, extracellular activities via TLR2, TLR4, and/or receptor for advanced glycation end products (RAGE).[64,65] However, the dependence of HMGB1 protein's proinflammatory cytokine effect on TLR2 and/or TLR4 could not be consistently demonstrated.[66] Recently, Rouhiainen et al.[67] reported that highly purified rhHMGB1 protein produced in a baculovirus system was essentially free of cytokine-inducing effect. In contrast, HMGB1 protein binds tightly to proinflammatory bacterial products, and a lipid-soluble proinflammatory bacterial product could be separated from the rhHMGB1 protein produced in an *Escherichia coli* system by chloroform-methanol treatment. These results suggest that the proinflammatory cytokine effect of HMGB1 protein is at least in part due to bacterial product contamination.

2.1.4 SCOPE OF THE CHAPTER

Contaminating PAMPs undoubtedly have contributed to the reported promiscuity of TLR2 and TLR4. However, the extent of PAMP contamination in the reported putative TLR ligands is not clear. Thus, TLRs may be more specific than previously realized. The purposes of this chapter are (1) to analyze reasons for failure to recognize contaminating PAMPs being responsible for the putative TLR ligands, and (2) to provide strategies that may be used to avoid future designation of PAMP contamination as putative ligands of TLRs.

2.2 RULING OUT PAMP CONTAMINATION: POTENTIAL PITFALLS

The possibility of PAMP contamination, particularly LPS, in purified biological preparations is well recognized.[33,68] A number of strategies have been used to determine whether the observed cytokine effects are due to LPS contamination: (1) using the *Limulus* amebocyte lysate (LAL) assay to determine the presence or absence of LPS; (2) using heat sensitivity by boiling the preparation of interest for 15–60 minutes to denature proteins, but not LPS; (3) using polymyxin B to specifically inhibit the biological activities of LPS; (4) using other LPS inhibitors such as lipid IVa, LPS or lipid A from *Rhodopseudomonas spheroids* to competitively inhibit the biological activities of LPS; (5) using antibodies such as anti-HSP antibodies to differentiate the effects of LPS and HSPs; and (6) using proteinase K to degrade proteins, but not LPS.[5,34,69] However, the use of these strategies to determine whether the observed cytokine effects are due to LPS contamination is problematic.

Although the commercial LAL chromogenic assay (e.g., LAL QCL-1000, Cambrex Bio Science, Walkersville, MD) is sufficiently sensitive to quantify 0.1 EU (endotoxin unit)/ml (~ 20 pg/ml) of reference standard endotoxin, the LPS content in many reported preparations was not accurately determined. In addition, the amounts of LPS present in many preparations were not properly reported. As a result, the exact amount of LPS in the test sample was not clear.[34]

In 1993, Majde[68] recommended that a simple heat inactivation (i.e., boiling for at least 30 minutes) could be used to determine whether the observed effects of a protein preparation are due to LPS contamination. This has become the single most widely used standard to rule out LPS contamination as the cause of the observed effects.[70] This recommendation was based on the assumption that LPS was resistant to boiling. In fact, it is generally believed that the pyrogenic effect of LPS is resistant even to autoclaving, and that dry heat at 250 °C for 1–2 hours or at 180 °C for 4 hours is required to render a substance pyrogen free.[71] However, the biological effects of LPS, including its endotoxin activity (i.e., LAL gelating activity) and cytokine-inducing activity, have been repeatedly demonstrated to be sensitive to boiling.[46,47,70,72,73] LPS is a potent modulator of the innate immune system. LPS in pg/ml concentration is sufficient to induce proinflammatory cytokine release from macrophages. The amount of TNFα released by murine macrophages reaches a plateau at an LPS concentration of approximately 1 ng/ml. With an LPS concentration of less than 1 ng/ml, the effect of heat inactivation can be readily detected, even after boiling for only 15 minutes. In contrast, if an LPS concentration greater than 5 ng/ml is used, no difference is seen in the TNFα-inducing activity between control and boiled LPS, even after boiling for 30 minutes.[70] This is because sufficient residual LPS activity remains in the boiled LPS to induce the maximal release of TNFα from macrophages. Thus, one can conclude that LPS is heat sensitive or heat resistant, depending on the concentration of LPS used to induce cytokine release from macrophages. As most studies used an LPS concentration ranging from 10 ng/ml to 1 μg/ml to test LPS heat sensitivity, one would conclude that LPS is heat resistant. Conversely, the contaminating LPS concentration present in the test sample is usually less than 1 ng/ml, which will be readily shown to be heat sensitive. Therefore, when using LPS heat sensitivity as a criterion to determine whether the observed effect is a result of the contaminating LPS, it is imperative to compare the heat sensitivity of LPS at the same concentration that is present in the sample being tested.[70]

Polymyxin B is a polycationic antibiotic that binds to the anionic lipid A portion of LPS and inhibits its activity.[74] However, it has no effect on PAMPs other than LPS. When added to test samples, polymyxin B can be used to inhibit the effects of LPS. Thus, if the observed effects of a test sample are inhibitable by polymyxin B, then they are likely due to LPS contamination. On the other hand, if they are not inhibitable by polymyxin B, one cannot conclude that the observed effects are the genuine effects of the molecule of interest in the test sample, because they could be due to contamination by PAMPs other than LPS. The use of other LPS inhibitors, such as lipid IVa, LPS or lipid A from *Rhodopseudomonas spheroids*, suffers from similar limitations. It is of particular concern that none of the studies reporting endogenous ligands of TLR2 and TLR4 have made any attempts to rule out the possibility of contamination by TLR2 PAMPs.[34]

Inhibition of Hsp60-induced cytokine release from macrophages by anti-Hsp60 antibodies has been interpreted as evidence that the observed cytokine effects are due to Hsp60.[75,76] However, Habich et al.[52] and Osterloh et al.[77] have recently demonstrated that LPS binds to rhHsp60 and that it is the Hsp60-bound LPS, not Hsp60 itself, that is responsible for the observed cytokine effects of rhHsp60. Thus, it is possible that anti-Hsp60 antibodies may inhibit the observed cytokine effects by interfering with the interaction of Hsp60-bound LPS with the TLR4/MD2 receptor complex. There is also evidence that Hsp70, Hsp90, gp96, and HMGB1 protein bind LPS.[48,78–80]

Proteinase K degrades protein, but not LPS. It can be used to distinguish the effect of protein from that of LPS. However, in order to avoid the potential confounding effect of proteinase K, boiling was used in some studies to inactivate proteinase K.[76] Unfortunately, this last step also inactivates LPS, as mentioned above, thus defeating the original purpose.

The above approaches, used either singly or in combination, do not unequivocally rule out the possibility of whether the observed effects are due to contamination by PAMPs. In summary, the reasons for failure to recognize PAMP contaminants being responsible for the putative TLR ligands include (1) failure to use highly purified preparations free of PAMP contamination, (2) failure to recognize the heat sensitivity of LPS, and (3) failure to consider contaminant(s) other than LPS.[34]

2.3 APPROACHES FOR AVOIDING THE DESIGNATION OF PAMPS AS PUTATIVE TLR LIGANDS

The following strategies can be used to minimize the possibility of designating PAMPs, particularly LPS, as putative TLR ligands.

2.3.1 PURITY OF POTENTIAL LIGANDS

The importance of using highly purified preparations essentially free of contaminating PAMPs cannot be overemphasized. Not only recombinant products derived from bacterial systems but also purified natural preparations or recombinant products derived from insect or mammalian systems can be contaminated with PAMPs that are introduced during the purification processes.[34,68,69] When commercially available, the "highly purified, low-endotoxin" preparation should be used for the study. If the cytokine effects are present only in less purified preparations, but not in the highly purified, low-endotoxin preparation, then it is likely that the observed cytokine effects are due to LPS contamination.[46,47]

Before a preparation is used to determine whether it is a potential TLR ligand, its LPS content should be measured. The LAL assay is currently the gold standard for LPS detection and quantification. Gel clot, turbidometric, and chromogenic LAL assays are commercially available. Below is the procedure for measuring LPS (endotoxin) using a microplate LAL chromogenic assay from the LAL QCL-1000 (Cambrex Bio Science, Walkersville, MD).

Protocol 2.1: Measuring LPS by LAL Chromogenic Assay

1. Prepare a solution containing 1.0 EU/ml endotoxin by diluting 0.1 ml of the endotoxin stock solution with LAL reagent water, and vigorously vortex for at least 1 minute. From this solution, prepare three solutions each containing 0.5, 0.25, and 0.1 EU/ml endotoxin as described above. These solutions are used as standards.
2. Preequilibrate the microplate at 37 °C ± 1.0 °C in the heating block adapter.
3. While leaving the microplate at 37 °C ± 1.0 °C, dispense 50 µl of LAL reagent water (as a blank), standards, or samples into the appropriate microplate well.
4. At time T = 0, add 50 µl of LAL to each well. Briefly remove the microplate from the heating block adaptor, and repeatedly tap the side of the plate to facilitate mixing. Return the plate to the heating block adapter, and replace cover.
5. At T = 10 minutes, add 100 µl of substrate solution (prewarmed to 37 °C ± 1°C) to each well. Mix the well content as in step 4.
6. At T = 16 minutes, add 100 µl of stop reagent to each well. Mix the well content as in step 4.
7. Read the absorbance of each well at 405–410 nm with a microplate reader using distilled water to adjust the photometer to zero absorbance.
8. Subtract the mean absorbance value of the blank from the mean absorbance value of the standards and samples to calculate Δ absorbance. Determine the endotoxin concentration of the samples either graphically or by linear regression from the mean Δ absorbance of standards.

The amount of LPS present in the preparation should be expressed as EU or ng LPS per unit quantity (e.g., mg) of the potential ligand, allowing calculation of the exact amount of LPS present in test samples. When comparing the effect of the potential ligand with that of LPS, it should be performed at an LPS concentration identical to that present in the potential ligand.

2.3.2 Ruling Out LPS Contamination

Regardless of the content of LPS present in the preparation, before designating any molecule as a TLR4 ligand, it is essential to exclude the possibility of LPS contamination being responsible for the observed cytokine effects. This is because the endotoxin activity as measured by the LAL assay and the LPS cytokine-inducing effect are two different biological activities of LPS. Dissociation between these two activities has been previously reported.[81] A number of approaches may be used to rule out the possibility of LPS contamination being responsible for the observed cytokine effect.

2.3.2.1 Polymyxin B Neutralization

Polymyxin B has been used successfully to inhibit the biological activities of LPS by binding to its endotoxic moiety of lipid A.[74] This can be accomplished by adding polymyxin B to the incubation media prior to the addition of a test sample or LPS.[46,47] If the observed cytokine effects are completely inhibited by polymyxin B, then they are likely due to LPS contamination. On the other hand, if the observed cytokine effects are not inhibitable or are only partially inhibitable by polymyxin B, the possibility that the observed cytokine effects may be due to contamination by PAMPs other than LPS should be considered.

2.3.2.2 Heat Sensitivity

Due to the past misconception that LPS is heat resistant, heat sensitivity has been widely used in the past with misleading conclusions. Thus, the use of heat sensitivity as a criterion to rule out LPS contamination should be avoided. However, if it is used, it is essential to compare the heat sensitivity of LPS at the same concentration that is present in the sample being tested.

2.3.2.3 LPS Removal System

A number of commercially available endotoxin-removing systems, such as Detoxi-Gel (immobilized polymyxin B gel, Pierce Biotechnology, Rockford, IL) and EndoTrap (Cambrex Bio Science, Walkersville, MD), can be used to repurify the preparation of interest to ensure that contaminating LPS is not responsible for the observed cytokine effects. If, after repurification once or twice, the observed cytokine effects persist, then it is unlikely that they are due to LPS contamination. Below is the procedure for LPS removal using a Detoxi-Gel AffinityPak Column (1 ml) from Pierce.

Protocol 2.2: Removing LPS Using a Polymyxin B-Agarose Column

1. Regenerate the Detoxi-Gel Resin by washing the column with five resin-bed volume (i.e., 5 ml) of 1% sodium deoxycholate at room temperature, followed by 3–5 ml of pyrogen-free buffer or water to remove the detergent.
2. Equilibrate the Detoxi-Gel Resin with 3–5 ml of a suitable pyrogen-free buffer or water.
3. Apply the sample to the column. Add aliquots of pyrogen-free buffer or water, and collect the flow-through. With a gravity-flow column, the sample will begin to emerge from the column after void volume has been collected, which is about 90% of the bed volume. For greater efficiency, stop column flow after the sample has entered the resin bed, and incubate the column for 1 hour before collecting the sample.

One ml of Detoxi-Gel removes > 9,995 EU from a 5 ml challenge containing 10,000 EU of LPS. Some protein binds LPS tightly and may be retained by the immobilized polymyxin B gel along with LPS, resulting in a significant loss of the protein (e.g., up to 30% for rhHsp60).[47]

2.3.3   RULING OUT CONTAMINATION BY PAMPS OTHER THAN LPS

There are no readily available methods to rule out the possibility of contamination by PAMPs other than LPS. The immobilized polymyxin B gel column may be used to remove LPS as well as LPS-associated molecules of microbial origin.[47] One may also use the octyl-Sepharose column, a hydrophobic interaction chromatography,[82] to remove TLR2 ligands such as lipoteichoic acid and bacterial lipoproteins to exclude the possibility of their contamination. However, it is difficult to determine which contaminating TLR PAMPs are responsible for the observed cytokine effects. Ye and Gan[53] used the recombinant Hsp70 produced in *E. coli* deficient in flagellin to demonstrate that the IL-12-inducing effect of recombinant Hsp70 in Jurkat T cells was due to flagellin contamination.

In summary, with the current limitation in our ability to detect and eliminate contaminating PAMPs in purified biological preparations, extreme caution should be exercised before designating any endogenous molecules as TLR ligands.

NOTE

The views presented in this article are the author's and do not necessarily represent the views of the Department of Veterans Affairs.

REFERENCES

1. Medzhitov, R., P. Preston-Hurlburt, and C. A. Janeway Jr. A human homologue of the *Drosophila* Toll proteins signals activation of adaptive immunity. *Nature* 388, 394, 1997.
2. Takeda, K., T. Kaisho, and S. Akira. Toll-like receptors. *Annu. Rev. Immunol.* 21, 335, 2003.
3. Kaisho, T., and S. Akira. Toll-like receptor function and signaling. *Allergy Clin. Immunol.* 117, 979, 2006.
4. Berg, A. A. Endogenous ligands of Toll-like receptors: implications for regulating inflammatory and immune responses. *Trends Immunol.* 23, 509, 2002.
5. Tsan, M. F., and B. Gao. Endogenous ligands of Toll-like receptors. *J. Leukoc. Biol.* 76, 514, 2004.
6. Rifkin, I. R., E. A. Leaderbetter, L. Busconi, G. Viglanti, and A. Marshk-Rothstein. Toll-like receptors, endogenous ligands, and systemic autoimmune disease. *Immunol. Rev.* 204, 27, 2005.
7. Akira, S. TLR signaling. *Curr. Top. Microbiol. Immunol.* 311, 1, 2006.
8. O'Neill, L. A. J. 2006. How Toll-like receptors signal: What we know and what we don't know. *Curr. Opinion Immunol.* 18, 3, 2006.
9. Choe, J., M. S. Keller, and I. A. Wilson. Crystal structure of human Toll-like receptor 3 (TLR3) ectodomain. *Science* 309, 581, 2005.
10. Bell, J. K., J. Askins, P. R. Hall, D. R. Davies, and D. M. Segal. The dsRNA binding site of human Toll-like receptor 3. *Proc. Natl. Acad. Sci. USA* 103, 8792, 2006.

11. Bell, J. K., I. Botos, P. R. Hall, J. Askins, J. Shiloach, D. R. Davies, and D. M. Segal. 2006. The molecular structure of the TLR3 extracellular domain. *J. Endotoxin Res.* 12, 375, 2006.

12. Viriyakosol, S., P. S. Tobias, R. L. Kitchens, and T. N. Kirkland. MD-2 binds to bacterial lipopolysaccharide. *J. Biol. Chem.* 276, 38044, 2001.

13. Miyake, K. Roles for accessory molecules in microbial recognition by Toll-like receptors. *J. Endotoxin Res.* 12, 195, 2006.

14. Matsumoto, M., K. Funami, M. Tanabe, H. Oshiumi, M. Shingai, Y. Seto, A. Yamamoto, and T. Seya. Subcellular localization of Toll-like receptor 3 in human dendritic cells. *J. Immunol.* 171, 3154, 2003.

15. Latz, E., A. Schoenemeyer, A. Visintin, K. A. Fitzgerald, B. G. Monks, C. F. Knetter, E. Lien, N. J. Nilsen, T. Espevik, and D. T. Golenbock. TLR9 signals after translocating from the ER to CpG DNA in the lysosome. *Nat. Immunol.* 5, 190, 2004.

16. Sioud, M. Induction of inflammatory cytokines and interferon responses by double-stranded and single-stranded siRNA is sequence dependent and requires endosomal localization. *J. Mol. Biol.* 348, 1079, 2005.

17. Kariko, K., H. Ni, J. Capodici, M. Lamphier, and D. Weissman. mRNA is an endogenous ligand for Toll-like receptor 3. *J. Biol. Chem.* 279, 12542, 2004.

18. Sioud, M. Innate sensing of self and nonself RNAs by Toll-like receptors. *Trends Mol. Med.* 12, 167, 2006.

19. Takeuchi, O., K. Hoshino, T. Kawai, H. Sanjo, H. Takada, T. Ogawa, K. Takeda, and S. Akira. Differential roles of TLR2 and TLR4 in recognition of gram-negative and gram-positive bacterial cell wall components. *Immunity* 11, 443, 1999.

20. Takeuchi, O., S. Sato, T. Horiuchi, K. Hoshino, K. Takeda, Z. Dong, R. L. Modlin, and S. Akira. Cutting edge: role of Toll-like receptor 1 in mediating immune response to microbial proteins. *J. Immunol.* 169, 10, 2002.

21. Poltorak, A., X. He, I. Smirnova, M. Y. Liu, C. Van Huffel, X. Du, D. Birdwell, E. Alejos, M. Silva, C. Galanos, M. Freudenberg, P. Ricciardi-Castagnoli, B. Layton, and B. Beutler. sDefective LPS signaling in C3H/HeJ and C57BL/10ScCr mice: mutations in Tlr4 gene. *Science* 282, 2085, 1998.

22. Alexopoulou, L., A. C. Holt, R. Medzhitov, and R. A. Flavell. Recognition of double-stranded RNA and activation of NK-κB by Toll-like receptor 3. *Nature* 413, 732, 2001.

23. Lund, J. M., L. Alexopoulou, A. Sato, M Karow, N. C. Adams, N. W. Gale, A. Iwasaki, and R. A. Flavell. Recognition of single-stranded RNA viruses by Toll-like receptor 7. *Proc. Natl. Acad. Sci. USA* 101, 5598, 2004.

24. Heil, F., H. Hemmi, H. Hochrein, F. Ampenberger, C. Kirschning, S. Akira, G. Lipford, H. Wagner, and S. Bauer. Species-specific recognition of single-stranded RNA via Toll-like receptor 7 and 8. *Science* 303, 1525, 2004.

25. Hemmi H, Takeuchi O, Kawai T, Kaisho T, Sato S, Sanjo H, Matsumoto M, Hoshino K, Wagner H, Takeda K, Akira S. A Toll-like receptor recognizes bacterial DNA. *Nature* 408: 740, 2000.

26. Hayashi F, Smith KD, Ozinsky A, Hawn TR, Yi EC, Goodlett DR, Eng JK, Akira S, Underhill DM, Aderem A. The innate immune response to bacterial flagellin is mediated by Toll-like receptor 5. *Nature* 410: 1099, 2001.

27. Yarovinsky F, Zhang D, Andersen JF, Bannenberg GL, Serhan CN, Hayden MS, Hieny S, Sutterwala FS, Flavell RA, Ghosh S, Sher A. TLR11 activation of dendritic cells by a protozoan profilin-like protein. *Science* 308: 1626, 2005.

28. Kurt-Jones, E. A., L. Popova, L. Kwinn, L. M. Haynes, L. P. Jones, R. A. Tripp, E. E. Walsh, M. W. Freeman, D. T. Golenbock, L. J. Anderson, and R. W. Finberg. Pattern recognition receptors TLR4 and CD14 mediate response to respiratory syncythia virus. *Nat. Immunol.* 1, 398, 2000.

29. Hietaranta, A., H. Mustonen, P. Puolakkainen, R. Haapiainen, and M. Kemppainen. Proinflammatory effects of pancreatic elastase are mediated through TLR4 and NF-κB. *Biochem. Biophys. Res. Commun.* 323, 192, 2004.

30. Miller, Y. I., S. Viriyakosol, C. J. Binder, J. R. Feramisco, T. N. Kirkland, and J. L. Witztum. Minimally modified LDL binds to CD14, induces macrophage spreading via TLR4/MD-2, and inhibits phagocytosis of apoptotic cells. *J. Biol. Chem.* 278, 1561, 2003.

31. Johnson, G. B., G. J. Brunn, Y. Kodaira, and J. L. Platt. Receptor-mediated monitoring of tissue well-being via detection of soluble heparan sulfate by Toll-like receptor 4. *J. Immunol.* 168, 5233, 2002.

32. Termeer, C., F. Benedix, J. Sleeman, C. Fieber, U. Voith, T. Ahrens, K. Miyake, M. Freudenberg, C. Galanos, and J. C. Simon. Oligosaccharides of hyaluronan activate dendritic cells via Toll-like receptor 4. *J. Exp. Med.* 195, 99, 2002.

33. Tsan, M. F., and B. Gao. Cytokine function of heat shock proteins. *Am. J. Physiol. Cell Physiol.* 286, C739, 2004.

34. Tsan, M. F., and B. Gao. Review: pathogen-associated molecular pattern contamination as putative endogenous ligands of Toll-like receptors. *J. Endotoxin. Res.* 13, 1, 2007.

35. Matzinger, P. The danger model: a renewed sense of self. *Science* 296, 301, 2002.

36. Janeway, C. A., Jr. The immune system evolved to discriminate infectious nonself from noninfectious self. *Immunol. Today* 13, 11, 1992.

37. Johnson, G. B., G. J. Brunn, A. H. Tang, and J. L. Platt. Evolutionary clues to the functions of the Toll-like family as surveillance receptors. *Trend Immunol.* 24, 19, 2003.

38. Seong, S. Y., and P. Matzinger. Hydrophobicity: an ancient damage-associated molecular pattern that initiates innate immune responses. *Nat. Rev.* 4, 469, 2004.

39. Bianchi, M. E. DAMPs, PAMPs and alarmins: all we need to know about danger. *J. Leukoc. Biol.* 81, 1, 2007.

40. Vabulas, R. M., P. Ahmad-Nejad, C. da Costa, T. Miethke, C. J. Kirschning, H. Häcker, and H. Wagner. Endocytosed HSP60s use Toll-like receptor 2 (TLR2) and TLR4 to activate the Toll/interleukin-1 receptor signaling pathway in innate immune cells. *J. Biol. Chem.* 276, 31332, 2001.

41. Asea, A., M. Rehli, E. Kabingu, J. A. Boch, O. Baré, P. E. Auron, M. A. Stevenson, and S. K. Calderwood. Novel signal transduction pathway utilized by extracellular Hsp70: role of Toll-like receptor (TLR) 2 and TLR4. *J. Biol. Chem.* 277, 15028, 2002.

42. Vabulas, R. M., P. Ahmad-Nejad, S. Ghose, C. J. Kirschning, R. D. Issels, and H. Wagner. Hsp70 as endogenous stimulus of the Toll/interleukin-1 receptor signal pathway. *J. Biol. Chem.* 277, 15107, 2002.

43. Vabulas, R. M., S. Braedel, N. Hilf, H. Singh-Jasuja, S. Herter, P. Ahmad-Nejad, C. J. Kirschning, C. da Costa, H-G. Rammensee, H. Wagner, and H. Schild. The endoplasmic reticulum-resident heat shock protein Gp96 activates dendritic cells via the Toll-like receptor 2/4 pathway. *J. Biol. Chem.* 277, 20847, 2002.

44. Wallin, R. P., A. Lundqvist, S. H. More, A. von Bonin, R. Kiessling, and H. G. Ljunggren. Heat-shock proteins as activators of the innate immune system. *Trends Immunol.* 23, 130, 2002.

45. Bausinger, H., D. Lipsker, U. Ziylan, S. Manie, J-P. Briand, J-P. Cazenave, S. Muller, J-F. Haeuw, C. Ravanat, H. de la Salle, and D. Hanau. Endotoxin-free heat shock protein 70 fails to induce APC activation. *Eur. J. Immunol.* 32, 3708, 2002.

46. Gao, B., and M. F. Tsan. Endotoxin contamination in recombinant human heat shock protein 70 (Hsp70) preparation is responsible for the induction of tumor necrosis factor α release by murine macrophages. *J. Biol. Chem.* 278, 174, 2003.

47. Gao, B., and M. F. Tsan. Recombinant human heat shock protein 60 does not induce the release of tumor necrosis factor α from murine macrophages. *J. Biol. Chem.* 278, 22523, 2003.

48. Reed, R. C., B. Berwin, J. P. Baker, and C. V. Nicchitta. GRP94/gp96 elicits ERK activation in murine macrophages: a role for endotoxin contamination in NFκB activation and nitric oxide production. *J. Biol. Chem.* 278, 31853, 2003.

49. Chen, K., J. Lu, L. Wang, and Y-H. Gan. Mycobacterial heat shock protein 65 enhances antigen cross-presentation in dendritic cells independent of Toll-like receptor 4 signaling. *J. Leukoc. Biol.* 75, 260, 2004.

50. Gao, B., and M. F. Tsan. Induction of cytokines by heat shock proteins and endotoxin in murine macrophages. *Biochem. Biophys. Res. Commun.* 317, 1149, 2004.

51. Osterloh, A., F. Meier-Stiegen, A. Veit, B. Fleischer, A. von Bonin, and M. Breloer. Lipopolysaccharide-free heat shock protein 60 activates T cells. *J. Biol. Chem.* 279, 47906, 2004.

52. Habich, C., K. Kempe, R. van Der Zee, R. Rumenapf, H. Akiyama, H. Kolb, and V. Burkart. Heat shock protein 60: specific binding of lipopolysaccharide. *J. Immunol.* 174, 1298, 2005.

53. Ye, Z., and Y-H. Gan. Flagellin contamination of recombinant heat shock protein 70 is responsible for its activity on T cells. *J. Biol. Chem.* 282, 4479, 2007.

54. Jaffray, C., C. Mendez, W. Denham, G. Carter, and J. Norman. Specific pancreatic enzymes activate macrophages to produce tumor necrosis factor-α: role of nuclear factor κB and inhibitory κB proteins. *J. Gastrointest. Surg.* 4, 370, 2000.

55. Geisler, F., H. Algul, M. Riemann, and R. M. Schmid. Questioning current concepts in acute pancreatitis: endotoxin contamination of porcine pancreatic elastase is responsible for experimental pancreatitis-associated distant organ failure. *J. Immunol.* 174, 6431, 2005.

56. Labarrere, C. A., and G. P. Zaloga. C-reactive protein: from innocent bystander to pivotal mediator of atherosclerosis. *Am. J. Med.* 117, 499, 2004.

57. Ballou, S. P., and G. Lozanski. Induction of inflammatory cytokine release from cultured human monocytes by C-reactive protein. *Cytokine* 4, 361, 1992.

58. Pasceri, V., J. T. Willerson, and E. T. H. Yeh. Direct proinflammatory effect of C-reactive protein on human endothelial cells. *Circulation* 102, 2165, 2000.

59. Bisoendial, R. J., J. J. P. Kastelein, J. H. M. Levels, I J. Zwaginga, B. van den Bogaard, P H. Reitsma, J. C. M. Meijers, D. Hartman, M. Levi, and E. S. G. Stroes. Activation of inflammation and coagulation after infusion of C-reactive protein in humans. *Cir. Res.* 96, 714, 2005.

60. Pepys, M. B., P. N. Hawkins, M. C. Kahan, G. A. Tennent, J. R. Gallimore, D. Graham, C. A. Sabin, A. Zychlinsky, and J. de Diego. Proinflammatory effects of bacterial recombinant human C-reactive protein are caused by contamination with bacterial products, not by C-reactive protein itself. *Cir. Res.* 97, e97, 2005.

61. Guillot, L., V. Balloy, F. X. McCormack, D. T. Golenbock, M. Chignard, and M. Si-Tahar. Cutting edge: the immunostimulatory activity of the lung surfactant protein-A involves Toll-like receptor 4. *J. Immunol.* 168, 5989, 2002.

62. Yamada, C., H. Sano, T. Shimizuet, H. Mitsuzawa, C. Nishitani, T. Himi, and Y. Kuroki. Surfactant protein A directly interacts with TLR4 and MD-2 and regulates inflammatory cellular response: importance of supratrimeric oligomerization. *J. Biol. Chem.* 281, 21771, 2006.

63. Bustin, M. Regulation of DNA-dependent activities by the functional motifs of the high-mobility group chromosomal proteins. *Mol. Cell. Biol.* 19, 5237, 1999.

64. Park, J. S., D. Svetkauskaite, Q. He, J-Y. Kim, D. Strassheim, A. Ishizaka, and E. Abraham. Involvement of Toll-like receptors 2 and 4 in cellular activation by high mobility group box 1 protein. *J. Biol. Chem.* 279, 7370, 2004.

65. Yang, H., H. Wang, C. J. Czura, and K. J. Tracey. The cytokine activity of HMGB1. *J. Leukoc. Biol.* 78, 1, 2005.

66. Yang, D., Q. Chen, H. Yang, K. J. Tracey, M. Bustin, and J. J. Oppenheim. High mobility group box 1 (HMGB1) protein induces the migration and activation of human dendritic cells and acts as an alarmin. *J. Leukoc. Biol.* 81, 59, 2007.
67. Rouhiainen, A., S. Tumova, L. Valmu, N. Kalkkinen, and H. Rauvala. Analysis of proinflammatory activity of highly purified eukaryotic recombinant HMGB1 (amphoterin). *J. Leukoc. Biol.* 81, 49, 2007.
68. Majde, J. A. Microbial cell-wall contaminants in peptides: a potential source of physiological artifacts. *Peptides* 14, 629, 1993.
69. Wakelin SJ, Sabroe I, Gregory CD, Poxton IR, Forsythe JL, Garden OJ, Howie SE. "Dirty little secrets"-endotoxin contamination of recombinant proteins. *Immunol Lett* 106: 1, 2006.
70. Gao, B., Y. Wang, and M. F. Tsan. The heat sensitivity of cytokine-inducing effect of lipopolysaccharide. *J. Leukoc. Biol.* 80, 359, 2006.
71. Sharma, S. K. Endotoxin detection and elimination in biotechnology. *Biotechnol. Appl. Biochem.* 8, 5, 1986.
72. Piotrowicz, B. I., and A. C. McCartney. Effect of heat on endotoxin in plasma and in pyrogen-free water, as measured in the *Limulus* amoebocyte lysate assay. *Can. J. Microbiol.* 32, 763, 1986.
73. Fujii, S., M. Takai, and T. Maki. Wet heat inactivation of lipopolysaccharide from *E. coli* serotype O55:B5. *PDA J. Pharmaceutical Sci. Technol.* 56, 220, 2002.
74. Morrison, D. C., and D. M. Jacobs. Binding of polymyxin B to the lipid A portion of bacterial lipopolysaccharide. *Immunochemistry* 13, 813, 1976.
75. Retzlaff, C., Y. Yamamoto, P. S. Hoffman, H. Friedman, and T. W. Klein. Bacterial heat shock proteins directly induce cytokine mRNA and interleukin-1 secretion in macrophage cultures. *Infect. Immun.* 62, 5689, 1994.
76. Bulut, Y., K. S. Michelsen, L. Hayrapetian, Y. Naiki, R. Spallek, M. Singh, and M. Arditi. *Mycobacterium tuberculosis* heat shock proteins use diverse TLR pathways to activate proinflammatory signals. *J. Biol. Chem.* 280, 20961, 2005.
77. Osterloh, A., U. Kalinke, S. Weiss, B. Fleischer, and M. Breloer. Synergistic and differential modulation of immune responses by HSP60 and LPS. *J. Biol. Chem.* 282, 4669, 2007.
78. Byrd, C. A., W. Bornmann, H. Erdjument-Bromage, P. Tempst, N. Pavletich, N. Rosen, C. F. Nathan, and A. Ding. Heat shock protein 90 mediates macrophage activation by taxol and bacterial lipopolysaccharide. *Proc. Natl. Acad. Sci. USA* 96, 5645, 1999.
79. Triantafilou, K., M. Triantafilou, and R. L. Dedrick. A CD14-independent LPS receptor cluster. *Nat. Immunol.* 2, 338, 2001.
80. Li, J., H. Wang, J. M. Mason, J. Levine, M. Yu, L. Ulloa, C. J. Czura, K. J. Tracey, and H. Yang. Recombinant HMGB1 with cytokine-stimulating activity. *J. Immunol. Methods* 289, 211, 2004.
81. Tian, L., J. E. White, H. Y. Lin, V. S. Haran, J. Sacco, G. Chikkappa, F. B. Davis, P. J. Davis, and M-F. Tsan. Induction of MnSOD in human monocytes without inflammatory cytokine production by a mutant endotoxin. *Am. J. Physiol. Cell. Physiol.* 275, C740, 1998.
82. Fisher, W. One-step purification of bacterial lipid macroamphiphiles by hydrophobic interaction chromatography. *Anal. Biochem.* 194, 353, 1991.

3 Analysis of TLR Expression, Regulation, and Signaling

Hui Xiao, Xiaoxia Li, and Derek W. Abbott

CONTENTS

3.1 INTRODUCTION

The innate immune system recognizes and responds to pathogenic organisms. In doing so, this system is responsible for initiating a cytokine response designed to tailor the adaptive immune system to eradicate the offending organism. Because this initial cytokine release must be tightly regulated, signal transduction pathways leading to this cytokine release are highly coordinated. This coordination begins at the cell surface with the initial recognition of pathogens by Toll-like receptors (TLRs). TLRs recognize components of bacteria, fungi, or viruses (collectively called pathogen-associated molecular patterns, or PAMPs) and play a major role in host defense against infection.[1] The majority of the TLR family members are abundantly expressed in monocytes, macrophages, and dendritic cells.[2] These receptors activate a highly conserved signaling network and ultimately lead to the activation of a variety of transcription factors, including NF-κB, ATF2, c-JUN, and IRF3/7. These transcription factors then act in synergy to induce the expression of hundreds of cellular genes.[3]

This gene induction is cell-type specific and determines the adaptive immune response. For example, monocytes and macrophages produce proinflammatory cytokines such as IL-1, TNFα, and interferons to initiate an acute inflammatory response,[4] while TLR engagement in dendritic cells induces the expression of co-stimulatory molecules that facilitate dendritic cell maturation and antigen presentation ability.[5] TLRs are also expressed in a subset of B and T cells, and recent work suggests that lymphocytic TLR signaling is essential to regulate lymphocyte activation and proliferation, antibody production, and antigen presentation.[6,7] Lastly, a subset of TLRs (i.e., TLR2, TLR4, and TLR9) is expressed by endothelial cells and by certain epithelial cells, where they appear to play a role in maintaining mucosal homeostasis. Thus, TLR signaling in endothelial cells or gastrointestinal epithelial cells may contribute to the pathogenesis of atherosclerosis and colitis, respectively.[8,9]

Some TLRs within the TLR family also show a cell-type-specific expression pattern. TLR3 is mainly expressed in dendritic cells, and can be induced in macrophages upon LPS stimulation.[10] TLR5 is highly expressed on the basolateral surface of the GI epithelium,[11] while TLR11 is expressed in the kidneys and liver.[12] In addition, TLR7 and TLR9, but not the other TLRs, are present in plasmocytoid dendritic cells (pDCs), and cause robust secretion of type I interferons aimed at inhibiting viral infection.[13,14] This cell-type-specific expression of TLRs argues for the existence of strict control of TLR expression. Very little is known about the mechanisms underlying this tissue-specific expression; however, it is tempting to speculate that the cytokines that regulate immune development and differentiation may play a role in this process. In this regard, proinflammatory cytokines such as TNFα and IFNγ, as well as bacterial or viral infections, have been shown to induce the expression of TLR2 and TLR3, but not the other TLRs.[8,15,16] These studies highlight the likely role of transcriptional factors NF-κB, STATs, and IRF3/7 in the regulation of TLR expression.

An additional level of regulation of the TLRs lies in their intracellular location. While TLRs 2, 4, and 6 are plasma membrane–localized receptors, TLRs 3, 7, and 9 are located intracellularly within endosomal compartments,[17,18] This cellular localization is required for TLRs 3 and 9 to recognize a viral infection, while at the same time not allowing the inadvertent recognition of host nucleic acids. This allows a degree of specificity to the differentiation of host versus pathogen.[19] Lastly, TLR9 is localized in the endoplasmic reticulum in resting cells, but translocates to the endosome when CpG or viral DNA is presented.[18,20] In fact, pDCs produce much more type I interferons than conventional DCs (cDCs) because CpG is translocated into the endosome in pDCs, while it is translocated mainly into the lysosome in cDCs.[21] This type of regulation of ligand and receptor translocation is crucial for appropriate TLR signaling and again illustrates an additional level of regulation by which host and pathogen can be distinguished.

Despite the complexity of both the expression and the localization of the TLRs, a basic mechanism by which the TLRs transmit signaling information has emerged. Upon ligand engagement, the TLRs oligomerize. With the exception of TLR3, these oligomerized TLRs recruit MyD88 to induce a highly conserved signaling pathway leading to NF-κB activation.[22] This signaling pathway involves dynamic complex formation and a cascade of kinase activation featuring four key mechanisms by which TLR signaling is achieved[22–24] (see Figure 3.1). Thus, immediately after the engagement of TLR, complex I (consisting of TLR-MyD88-IRAK4-IRAK1-TRAF6) is formed. In this complex, the death domain–containing serine/threonine kinases

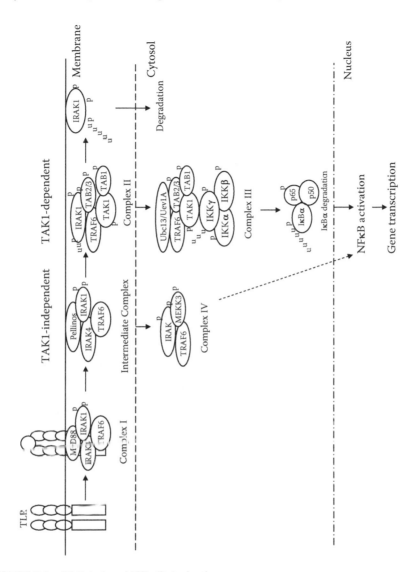

FIGURE 3.1   TLR-induced NF-κB Activation

Note: Upon TLR ligand stimulation, adapter molecules, such as MyD88 and/or Trif, associate with the engaged receptor and, in turn, recruit IRAK1, IRAK4, and TRAF6, resulting in the formation of the receptor complex (complex I). Upon the formation of complex I, IRAK4 is activated, leading to the hyperphosphorylation of IRAK1, which creates an interface for the interaction of Pellino proteins with the IRAK1-IRAK4-TRAF6 complex (intermediate complex). IRAK1 bound to TRAF6 then leaves the intermediate complex and binds to TAK1-TAB1-TAB2/3 to form complex II on the plasma or endosomal membrane. In complex II, IRAK1 is ubiquitinated and degraded, resulting in the translocation of the TRAF6-TAK1-TAB1-TAB2/3 complex (complex III) from the membrane to the cytosol, where TAK1 is activated. Subsequently, TAK1 activates IKKβ, which, in turn, activates NF-κB through the degradation of IκBα. In addition, intermediate complex can bind to MEKK3 to form complex IV in the cytosol. Complex IV can also mediate NF-κB activation through a process not involving IκBα degradation.

(i.e., IRAK4 and IRAK1) are activated. This activation leads to the phosphorylation of IRAK1, and phosphorylated IRAK1 bound to TRAF6 leaves complex I and binds to TAK1/TAB2/3 to form complex II on the plasma or endosomal membrane. In complex II, autoubiquitinated TRAF6 promotes the phosphorylation of TAK1/ TAB2/TAB3. At this point, hyperphosphorylated IRAK1 is polyubiquitinated, and subsequently degraded in a proteasome-dependent manner. This process facilitates the movement of the TRAF6-TAK1/TAB2/3 complex into the cytosol, where it binds to IKK to form complex III. In complex III, TAK1 phosphorylates the activation loop of IKKβ. Activated IKKβ then phosphorylates IκBα, ultimately targeting it for polyubiquitination and proteosomal degradation. Degradation of IκBα allows the NF-κB transcription factors to enter the nucleus and activate target genes Within this canonical activation pathway, additional, recently recognized proteins such as SIGIRR, Tollip, or Pellino proteins can also bind to various signaling complexes to either positively or negatively regulate NF-κB activation.[25,26] Furthermore, phosphorylation, ubiquitination, and sumoylation of the IKK scaffolding protein, NEMO, play a critical role in regulating IKKβ activity.[27] Lastly, in addition to the TAK1-dependent pathway described above, MEKK3 mediates NF-κB activation in IL-1 and TLR7 signaling.[28,29] However, the detailed mechanism by which this TAK1-independent pathway affects TLR signaling is only in its infancy and is in need of further investigation.

Given the precision with which TLRs must recognize their ligands and tailor the adaptive immune system response, it is not unexpected that their expression and activity are regulated at multiple levels. This chapter aims to serve as a synopsis by which this regulation can be studied. A set of methods designed to monitor the strength and duration of the TLR response is presented, and common artifacts associated with these methods as well as ways to circumvent them are discussed.

3.2   ANALYSIS OF TLR EXPRESSION

3.2.1   DETECTION OF TLR EXPRESSION AT THE MESSAGE LEVEL

Real-time polymerase chain reaction (PCR) and Northern blotting are two common procedures designed to measure TLR expression at the mRNA level. Compared to Northern blotting, real-time PCR is more quantitative, more sensitive, and more efficient, but also more expensive. Real-time PCR is highly favorable when limited amounts of cells or tissue are being studied. The protocol described below is based on SYBR green I dye chemistry, which is more versatile and money saving than the TaqMan Probe-based fluorogenic 5′ nuclease chemistry. Since SYBR green dye binds to any type of double-stranded DNA (dsDNA), DNase I treatment or uracil N-glycosylase should be performed to remove dsDNA from RNA or cDNA samples. By doing so prior to real-time PCR, the SYBR green dye will bind only to the dsDNA formed by the PCR reaction. Additionally, primers are designed to anneal to different exons such that contaminating genomic DNA is not amplified.

Protocol 3.1: Evaluation of Inducible TLR3 Expression by Real-Time PCR

1. Stimulate mouse peritoneal macrophages (8×10^6 cells plated in a 10 cm dish) with highly purified lipopolysaccharide (LPS, 1μg/ml; Invivogen, San Diego, CA) for 4 hours. Terminate the stimulation by washing the cells with 10 ml of cold PBS, and immediately add 1 ml of Trizol reagent (Sigma-Aldrich, St. Louis, Missouri) to the cells. Isolate total RNA using the protocol recommended by Sigma.
2. Reverse-transcribe the RNA using the SuperScript II RT kit (Invitrogen, Carlsbad, CA) according to the manufacturer's protocol. Dilute the cDNA tenfold with sterile water and store at −80 °C.
3. Design PCR primers using Primer3: WWW Primer Tool (http://biotools. umassmed.edu/bioapps/primer3_www.cgi).[30] Use the following settings:
Product size: minimum: 100 bp, optimal: 120 bp, maximum: 150 bp
Primer temperature: minimum: 60 °C, optimal: 63 °C, maximum: 65 °C
Primer size: minimum: 18 bp, optimal: 22 bp, maximum: 28 bp
Maximum 3′ stability: 7

   Use these settings while keeping the other settings at default. From the list of recommended primers, select the ones that fit two additional requirements:

   1. There is no more than two G's or C's in the terminal 5 bases at the 3′ of each primer.
   2. Both primers are located in different exons.

   We selected the following primer set for mouse *Tlr3* gene.
   Left primer: 5′ TAAAATCCTTGCGTTGCGAAGT 3′ (546)
   Right primer: 5′ TGTTCAAGAGGAGGGCGAATAA 3′ (691)
   Product size: 146 bp
4. Set up real-time PCR:
   Primers: 3 μl of each
   cDNA: 3 μl
   SYBR green PCR master mix (Applied Biosystems, Foster City, CA): 12.5 μl
   Sterile H_2O: 6.5 μl
   Run PCR in a 7300 or 7500 Real Time PCR System (Applied Biosystems) at the following settings:
   Step 1.　AmpErase: 50 °C for 2 minutes
   Step 2.　Denature: 95 °C for 10 minutes
   Step 3.　Denature: 95 °C for 15 seconds
   Step 4.　Anneal: 60 °C for 1 minute
   Run steps 3–4 40 times.
   After the amplification, run a dissociation program to check the melting curve.
   A housekeeping gene, such as actin, that is known not to be regulated by LPS should be analyzed in the same experiment to provide an internal control.

5. Calculate the relative abundance, that is, the ratio of the TLR3mRNA to actin mRNA using $\Delta\Delta$Ct method.

The PCR raw data can be analyzed by either relative abundance or standard curve. By plotting a standard curve for a target gene, the exact amount of TLR expression in each sample can be calculated. However, if expression of the whole set of TLRs is studied, as many as 13 standard curves need to be made. This is generally why relative abundance is chosen when multiple genes are analyzed from the same sample set. Relative expression of all the target genes can be calculated after normalization to an endogenous control, such as actin.

There is still a chance that even well-designed primers may also amplify other cDNAs present in the sample. If the dissociation curve shows more than one peak, or if more than one PCR product is detected by agarose gel electrophoresis, the PCR conditions will need to be optimized to amplify only one specific gene product. Furthermore, as an additional control, two pairs of primers that amplify different parts of a target gene can be used to demonstrate that they amplify the gene with the same relative abundance.

3.2.2 DETECTION OF TLR EXPRESSION AT THE PROTEIN LEVEL

While real-time PCR is designed to detect gene expression and regulation at the mRNA level, immunofluorescence staining and Western blotting are procedures used to detect TLR expression at the protein level. TLR2 is induced by LPS, IFNγ, or TNFα in a variety of cell types.[8,31] By applying the immunofluorescence protocol to the lung tissue, one can detect the expression of TLR2, or other TLRs, under LPS challenge, bacteria infection, or other inflammatory stimuli.

**Protocol 3.2: Detection of TLR2 Expression in
the Lung by Immunofluorescence**

1. Inject 400 µl of pure LPS (1 mg/ml in PBS buffer) or PBS intraperitoneally into 6–8-week-old mice. Sacrifice the animals 4 hours later, and transcardially perfuse them with PBS. Remove the lungs, and embed them in optimal cutting temperature medium (OCT) in a plastic base mold (Cardinal Health, Dublin, OH). Snap-freeze the lung tissue by floating the mold on liquid nitrogen for 1 minute or until the OCT freezes. Keep the frozen tissue at −80 °C.
2. Generate 5 µm thick cryosections, place them on glass slides, and keep them frozen until use. Dry the cryosections for 1 hour at room temperature, and then fix the frozen tissue in cold acetone (−20 °C) for 10 minutes in a plastic copier jar. Wash the slides with PBS three times (5 minutes each time). Remove the water from the slide carefully while keeping the tissue humidified, and then draw a circle around the tissue with a waterproof mark pen (Pap-pen) to keep the liquid covering the tissue during the following incubation steps.

3. Cover the cryosections with 100–200 μl of blocking buffer (1% bovine serum albumin [BSA] in PBS) and incubate for 1 hour at 37 °C. To prevent drying out of the tissue, the slides should be kept in a slide box or a Petri dish with water-saturated paper at the bottom of the dish. After blocking, remove the blocking buffer by aspiration with a pasture pipette. Wash the sections with PBS three times (5 minutes each time).

4. Cover the sections with rabbit anti TLR2 antibody (IMGENEX, San Diego, CA), diluted (1:500) in antibody dilution buffer (0.1% BSA, 0.05% sodium azide, 0.1% Triton-X, 0.04% EDTA, and 1× PBS), and incubate for 2 hours at 37 °C or overnight at 4 °C. Wash the slides three times with PBS (5 minutes each time). Keep the slides in the dark to prevent fluorescence bleaching.

5. Incubate the slides with Alexa fluor 488 conjugated goat anti-rabbit secondary antibody (Molecular Probes, Eugene, OR) in antibody dilution buffer (1:500) for 1 hour at 37 °C. Wash three times with PBS (5 minutes each time).

6. To stain endothelial cells in the lung tissue, repeat steps 4–5 with rat anti-CD31 (BD Biosciences, San Jose, CA) and Texas red conjugated goat anti-rat secondary antibody (Molecular Probes, Eugene, OR) in antibody dilution buffer.

7. Stain the nuclei by incubation with 4′,6-diamidino-2-phenylindole (DAPI) solution (diluted 1:15,000 in water) for 5 minutes at room temperature. Wash the slides with water extensively, immerse them in Vectashield Mounting medium without DAPI (Vector Laboratories, Burlingame, CA), and cover them with cover slips. Examine the TLR2 expression and localization under a fluorescence microscope.

The major caveat for immunofluorescence staining is that primary antibody may stain nonspecifically other cellular proteins. To exclude this possibility, the anti-TLR2 antibody should be preincubated with the peptide that was used to generate the antibody to saturate its antigen binding. The lack of staining with the peptide-neutralized antibody would indicate that this antibody stains only TLR2 in the tissue. Alternatively, tissues or cells, which don't express TLR2, can be used to test the antibody specificity as well. For example, LPS-treated lung tissue from TLR2-deficient mice could be tested, and the lack of staining would confirm the specificity of this antibody to TLR2.

In addition to endothelial lung cells, the expression of TLRs in other cell types within the lung tissue can be detected by co-staining for cytokeratin, F4/80, or CD11c, which are cellular markers for epithelial, macrophage, or dendritic cells, respectively. Therefore, a comprehensive expression pattern of TLRs can be obtained by a set of immunofluorescent stainings, and this helps us to understand how all the cells in an organ function cooperatively to regulate inflammatory responses.

3.3   ANALYSIS OF SIGNALING COMPLEXES

Signaling complexes assembled through protein-protein interaction are pivotal to transduce signal from membrane-bound TLR to the nucleus, where it induces gene expression. Co-immunoprecipitation is a most powerful approach to detect signaling complexes formed by multiple proteins in vivo.[23–26] Adaptors, such as MyD88, Trif, Mal, or TRAM, assemble on activated TLR receptors and form signaling complexes that mediate NF-κB or IRF activation.[1,22]

Protein-protein interaction is a dynamic process, and signaling complexes change throughout the duration of TLR signaling. To elucidate the dynamic complex formation, co-immunoprecipitations against various components of the signaling complexes can be performed. By performing anti-IL-1R, anti-IRAK (Santa Cruz Biotechnology, Santa Cruz, CA), and anti-TRAF6 (Santa Cruz Biotechnology) co-immunoprecipitations, we observed that IL-1R-MyD88-IRAK4-IRAK1-TRAF6 complexes can be detected as early as 1 minute following IL-1 stimulation.[23–26] The IRAK1-TRAF6 complex then dissociates from the receptor complex. It is important to note that the TLR signaling is transduced from the membrane to the cytosol, and finally to the nucleus, and that co-immunoprecipitation can also detect the localization of signaling complexes if biochemical fractionation is conducted. A widely used biochemical fractionation protocol can be employed to prepare protein lysates of cytosol, nuclei, and cell membranes before and after TLR stimulation. Co-immunoprecipitation studies of subcellular fractions indicate that IRAK1-TRAF6-TAK1 complexes are initially formed on the plasma membrane, and then TAK1 together with TRAF6 translocate into the cytosol and phosphorylate IKK.[24]

Protocol 3.3: Detection of Complex Formation by Co-immunoprecipitation

1. Stimulate cells (1.5×10^7) of choice with respective TLR ligands for 0, 5, 15, 30, or 60 minutes.
2. Terminate this stimulation by removing the medium and washing the cells with 15 ml of cold PBS. Add 1 ml of ice-cold lysis buffer (20 mM HEPES [pH 7.4], 150 mM NaCl, 1.5 mM $MgCl_2$, 2 mM EDTA, 2 mM DTT, and 1% Triton X-100) supplemented with 1 mM phenylmethylsulfonyl fluoride (PMSF), 1 mM sodium fluoride (NaF), 1 mM sodium orthovanadate (Na_3VO_4), and protease inhibitors (Complete, from Roche Diagnostics, Basel, Switzerland). Detach the cells with a cell lifter (Fisher Scientific, Waltham, MA), and transfer cell suspension into a 1.5 ml tube.
3. Lyse the cells by pipetting the cell suspension approximately 10 times, followed by incubation on ice for 20 minutes to solubilize proteins. Centrifuge the cell lysates at 13,000 rpm for 10 minutes at 4 °C, discard the pellet, and transfer the supernatant into a fresh 1.5 ml microfuge tube.
4. Preclear the lysates with empty protein A beads loaded with nonspecific antibody: transfer 40 μl of 50% slurry of beads (Amersham Biosciences, Piscataway, NJ) into a 1.5 ml tube. Wash the beads with 1 ml of PBS by centrifugation at 8,000 rpm for 1 minute and removing the wash buffer by aspiration. Incubate the lysates with 20 μl of washed protein A Sepharose beads and 1 μg of nonspecific antibody for 1 hour at 4 °C on a rotator. Fol-

lowing centrifugation at 8,000 rpm for 1 minute, transfer these precleared cell lysates into a fresh 1.5 ml tube.

5. Immunoprecipitate precleared cell lysates by adding 2 µg of respective anti-TLR for 4 hours at 4 °C, followed by incubation with 20 µl of washed protein A agarose beads for another 1 hour. Collect the beads by centrifugation at 8,000 rpm for 1 minute. Carefully remove the supernatant by aspiration. Wash the beads with 1 ml of lysis buffer by spinning down the beads at 8,000 rpm for 1 minute. Remove the wash buffer by aspiration, and repeat the wash four times.

6. Solubilize the immunoprecipitates by adding 20 µl of 2× Laemmli SDS-PAGE sample buffer. Resolve proteins by 10% SDS polyacrylamide gel electrophoresis (SDS-PAGE), and transfer them onto a PVDF membrane (Millipore, Billerica, MA). Immunoblot the PVDF membrane with anti-MyD88 antibody (Stressgene, Victoria, BC, Canada), anti-IRAK4 antibody (Cell Signaling Technology, Danvers, MA), anti-IRAK1 antibody (Santa Cruz Biotechnology, Santa Cruz, CA), or anti-TRAF6 antibody. The same PVDF membrane can be stripped and reprobed with different antibodies. Thus, following enhanced chemiluminescence (ECL), incubate the membrane in stripping buffer (62.5 mM Tris [pH 8.0], 2% SDS, and 0.1 M β-mercaptoethanol) for 30 minutes at 50 °C, and then wash the membrane extensively with TBST buffer (20 mM Tris-HCl [pH 7.5], 150 mM NaCl, and 0.1% Tween 20).

When assessing co-immunoprecipitation experiments, one needs to be cognizant of common artifacts. Nonspecific association of proteins to IgG or Sepharose beads is a common cause of false positives in co-immunoprecipitation assays. Preclearing cell lysates by preincubation with Sepharose beads helps to reduce nonspecific protein binding. However, if nonspecific binding is still present, it is highly recommended to wash the immunoprecipitated beads under a more stringent condition (i.e., by increasing the concentration of NaCl to 350–500 mM or by including 0.1% SDS in the wash buffer). While antibody specificity is a key for a successful co-immunoprecipitation assay, the presence of stable protein complexes formed in the lysates is equally crucial. To detect relatively weak or unstable protein complexes, irreversible cross-linking reagents can be utilized, although artifactual binding can also occur with these reagents.

Other approaches have been designed to examine protein-protein interactions. GST pull-down assays are designed to study the interaction of a GST-tagged protein, which in general is expressed and purified from bacteria *E. coli*, and a candidate protein, which is expressed and purified from *E. coli* or synthesized by an in vitro transcription-translation system, in an in vitro binding assay. This in vitro pull-down assay is useful to detect interaction between two proteins by direct contact, rather than indirect contact. Co-localization studies based on immunofluorescence staining or fluorescence resonance energy transfer (FRET) can detect protein-protein interactions in vivo and can also uncover where in the cell the protein-protein interaction occurs.[32,33]

3.4   ANALYSIS OF IRAK1 MODIFICATION AND DEGRADATION

Following TLR (or IL-1) stimulation, IRAK1 is multiphosphorylated and undergoes K48-linked polyubiquitination. Upon polyubiquitination, IRAK1 is degraded by the proteosome.[34] These extensive posttranslational modifications and the ultimate degradation of IRAK1 can be detected by Western blotting—first by the appearance of slower migrating bands, and then by their disappearance.[23–26,35] The apparent molecular weight of unmodified full-length IRAK1 protein is approximately 80 kD, and our labs have detected IRAK1 protein bands present as a ladder ranging from 80 kD to 250 kD. The modified IRAK1 protein bands start to appear as early as 5 minutes after Pam_3Cys_4 or IL-1 stimulation on anti-IRAK1 Western blots as bands between 80 to 130 kD mainly attributable to the hyperphosphorylation of IRAK1.[23–26] After 15 minutes, polyubiquitinated IRAK1 becomes more evident as bands over 200 kD. During this time, the amount of 80 kD IRAK1 protein decreases, owing to the posttranslational modifications and the proteasome-mediated IRAK1 degradation. After 4 hours following stimulation, only a very small amount of IRAK1 can be detected. As IRAK1 is a positive regulator of TLR signaling, its degradation deceases the overall TLR signaling. Surprisingly (and owing to the complexity of signaling systems), our recent studies suggest that the degradation of IRAK1 is also necessary to promote NF-κB activation, and thus can also serve as a marker of TLR activation.[28]

To add to the complexity, IRAK1 modifications and degradation also show cell-type and/or signaling specificity. Pam_3Cys_4 and LPS, but not LTA, can induce IRAK degradation in monocytic cell lines THP1 and Mono Mac 6.[36,37] Furthermore, MALP-1, LPS, and R848, but not CpG, induce IRAK1 degradation in peritoneal macrophages and in fibroblasts.[38–40] Given these cell-type specificities, while IRAK1 modification and degradation comprise a useful marker for TLR stimulation, each cell type must be individually tested.

Protocol 3.4: Examination of IRAK1 Modification and Degradation

1. Grow a suspension of 10^7 THP-1 cells in 25 ml of Dulbecco's modified Eagle's medium (DMEM; Invitrogen, Carlsbad, CA) supplemented with 10% fetal bovine serum at 37 °C in a 5% CO_2 environment. Stimulate the cells with 100 ng/ml of Pam_3Cys_4 (Invivogen, San Diego, CA) for 0, 5, 15, 30, or 60 minutes.
2. After the allotted time has elapsed, spin the cells at 1,000 rpm for 3 minutes, remove the medium by aspiration, and wash the cells gently with 10 ml of PBS containing 10 nM calyculin (Millipore, Billerica, MA). Spin the cells again, and lyse in 200 µl of lysis buffer (20 mM HEPES [pH 7.4], 150 mM NaCl, 2 mM EDTA, 1% Triton X-100, and 0.1% SDS) supplemented with 1 mM PMSF, 1 mM NaF, 10 nM calyculin, 1 mM Na_3VO_4, and protease inhibitors (Complete, from Roche Diagnostics, Basel, Switzerland).

3. Vortex the cells vigorously, and incubate them on ice for 20 minutes. Spin down the insoluble materials at 13,000 rpm for 10 minutes at 4 °C. Transfer the cell lysates into 1.5 ml microfuge tubes.

4. Measure the protein concentration in cell lysates with the BCA protein assay kit (Pierce Biotechnology, Rockford, IL). Take 100 µg of total protein from each sample, and mix with an equal volume of 2× Laemmli SDS-PAGE sample buffer. Heat the lysates at 95 °C for 5 minutes, and place on ice.

5. Resolve proteins by SDS-PAGE on an 8% polyacrylamide gel, and transfer them onto a 45 µM pore PVDF membrane (Milipore) using standard transfer techniques.

6. Block the PVDF membrane with 5% milk in TBST (20 mM Tris-HCl [pH 7.5], 150 mM NaCl, and 0.1% Tween 20) for 60 minutes. Wash the membrane twice (5 minutes per wash) with TBST buffer. Next, incubate the membrane with rabbit anti-IRAK1 antibody (Cell Signaling Technology, Danvers, MA) diluted in 5% milk–TBST buffer (1:1,000) overnight at 4 °C. Wash the membrane three times with TBST buffer vigorously (10 minutes for each wash).

7. Incubate the membrane with goat anti-rabbit secondary antibody conjugated with horseradish peroxidase (HRP; Jackson Labs, Bar Harbor, ME) diluted in 5% milk–TBST buffer (1:5,000) for 45 minutes at room temperature. Then, wash the membrane three times with TBST vigorously (10 minutes each). After the washings, incubate the membrane with ECL reagent (Amersham Biosciences, Piscataway, NJ) for 1 minute (use enough ECL to cover the blot completely), and detect the chemiluminescence signal by exposure to an X-ray film. A series of films exposed for different periods of time (0.5–5.0 min) should be used to obtain a linear range of signals.

3.5   ANALYSIS OF IRAK AND TAK1 ACTIVATION

The kinase activity of IRAK4 plays an essential role in the induction of proinflammatory gene expression in response to a subset of TLR ligands. Despite initial reports that IRAK4 kinase activity was nonessential for NF-κB signaling, recent work suggests that IRAK4's kinase activity helps to regulate both transcriptional activity and mRNA stability of induced genes.[41] In contrast, IRAK1 kinase activity appears to be nonessential for NF-κB signaling, but is instead implicated in the regulation of TLR7- and TLR9-induced type I interferon production in pDC through the proposed mechanism of direct phosphorylation of the transcription factor IRF7.[33] While both IRAK4 and IRAK1 kinases are activated by IL-1 induction, they appear to play a redundant role in NF-κB activation while playing a more specific role in IRF7 phosphorylation and mRNA stability.[23,33,41] In contrast, the downstream kinase, TAK1, is essential for optimal NF-κB signaling in response to an array of TLR and inflammatory agonists.[35]

To characterize the kinetics of the TLR signal transduction cascades, kinases need to be immunoprecipitated from mammalian cells upon ligand stimulation. Since immunoprecipitated IRAK or TAK1 is activated in vivo, the kinase assays can be used to study the strength, timing, and duration of signaling. While this is advantageous in that the immunoprecipitated kinases are more physiologically relevant, the experiments need to be carefully controlled to account for co-precipitation of irrelevant kinases, antibody specificity, and antibody affinity. For example, because IRAK4 and IRAK1 associate with each other in vivo,[23,24,40] IRAK4 is a contaminating kinase present in anti-IRAK1 immunoprecipitates. Likewise, upon stimulation, TAK1 associates with the IKK signalosome,[42] so IKKs can be present in TAK1 immunoprecipitates. These contaminating kinases could potentially give rise to false positive readouts of TLR signaling. One of the strategies to limit the impact of contaminating kinases is to increase the stringency of washes following immunoprecipitation. We have found that increasing the NaCl concentration in the wash buffer to 350–500 mM greatly decreases the presence of other kinases in the anti-IRAK1, anti-IRAK4, or anti-TAK1 immunoprecipitates.

Additionally, the use of a specific substrate can also allow a degree of specificity to the kinase reaction. Three kinds of substrates have been used for IRAK1 and IRAK4 kinase assays: myelin basic protein (MBP), IRAK1, and Pellino 2.[36,38,43] MBP is a widely used artificial substrate for kinase assays. MBP is convenient to use because of its commercial availability and its applicability to a broad range of kinases, especially when a kinase's substrate is unknown or unavailable. In addition, MBP is a highly basic protein containing a large number of surface-accessible phospho-acceptor sites that can be phosphorylated by many kinases. For these reasons, the use of MBP as a substrate does not control for co-precipitating kinases. The other substrates (i.e., kinase-dead IRAK1 and Pellino 2) allow better specificity of IRAK4 and IRAK1 detection, because no other kinases have been shown to be capable of phosphorylating these substrates. Therefore, kinase-dead IRAK1 and Pellino 2 are the specific readout for IRAK1 or IRAK4 activation by kinase assay. IRAK1 autophosphorylation is also often used as readout to examine IRAK1 activation in TLR signaling. Because both kinase and substrate are obtained by anti-IRAK1 immunoprecipitation under stringent washing conditions, the readout is relatively specific to IRAK1 activation. Lastly, in many cases, a truncated form of a protein that contains the phosphorylation site or even a peptide corresponding to the amino acids surrounding a phospho-acceptor site may be much easier to obtain and is, in many instances, available commercially. These types of substrates also decrease the effects of a co-precipitating kinase. Despite these caveats, when performed with the proper controls, kinase activity assays can be useful in studying the kinetics and strength of IL-1/TLR signaling.

Protocol 3.5: Determination of IRAK4 and TAK1 Activation

1. Stimulate cells (1.5×10^7) of choice with respective TLR ligands for 5 minutes (for IRAKs) or 10–15 minutes (for TAK1).
2. Remove the medium by aspiration, and wash the cells quickly with 10 ml of PBS. Scrap off cells in 1 ml of lysis buffer (50 mM Tris [pH 7.5], 150 mM NaCl, 1% NP-40, 1 mM β-glycerophosphate, 2.5 mM sodium pyrophosphate, 1 mM sodium fluoride, 1 mM EDTA, 1 mM ethylene glycol tetraace-

tic acid (EGTA), 1 mM PMSF, 1 mM leupeptin, 10 nM calyculin, and 1 mM sodium orthovanadate). Pipette the scraped cell suspension 10–15 times, and incubate it on ice for 20 minutes. Centrifuge the cell lysates at 13,000 rpm for 10 minutes at 4 °C. Discard the pellet, and transfer the supernatant into a fresh 1.5 ml tube.

3. Preclear the cell lysates by incubation with 20 µl of protein A or protein G Sepharose beads (40 µl of 50% slurry [Amersham Biosciences, Piscataway, NJ] washed with 1 ml of PBS twice) for 1 hour at 4 °C on a rotator. Immunoprecipitate the precleared cell lysates with 1.5 µg of anti-IRAK4 polyclonal serum (Zymed, San Francisco) or 1.5 µg anti-TAK1 antibody (Upstate Biotechnology, Lake Placid, NY) for 4 hours at 4 °C, followed by incubation with 20 µl of prewashed protein A Sepharose beads for 1 hour.

4. Collect the beads by centrifugation at 4,000 rpm for 1 minute, and carefully remove the supernatant by aspiration. Wash the beads with 1 ml of lysis buffer four times, followed by two more washes with 1 ml of kinase assay buffer (20 mM HEPES [pH 7.4], 20 mM $MgCl_2$, 1 mM DTT, 20mM β-glycerophosphate, 20 mM paranitrophenylphosphate, and 1 mM EDTA) supplemented with 1 mM NaF, 1 mM sodium vanadate, and protease inhibitors (Complete, from Roche Diagnostics, Basel, Switzerland).

5. Set up an in vitro kinase assay (total volume of 30 µl) containing the following:

> Beads containing anti-IRAK4 or anti-TAK1 immunoprecipitates (a control reaction contains immunoprecipates from untreated cells)
>
> 1 µg of purified kinase domain of IRAK1 (amino acids 182–546, and kinase-dead mutant, purified from SF9 insect cells) for IRAK4 assays, or bacterially produced, purified MKK6 (bacteria lack the upstream kinase of MKK6; thus, the bacterially produced MKK6 is dormant until activated) for TAK1 assays
>
> 20 µl of kinase assay buffer
>
> 100 µM adenosine triphosphate (ATP)
>
> 10 µCi of [γ-^{32}P]-labeled ATP (3,000 Ci/mmol; Amersham Biosciences, Piscataway, NJ)

Mix the solutions by gently tapping the tube, and incubate at 30 °C for 30 minutes. Stop the reaction by adding 30 µl of 2× Laemmli SDS-PAGE sample buffer, followed by heating for 5 minutes at 96 °C.

6. Briefly spin down the beads, and load proteins in the aqueous phase into a 10% SDS polyacrylamide gel. Resolve proteins by electrophoresis. Once the bromophenol blue dye migrates through 75% of the gel, stop the electrophoresis. Transfer the gel onto 3 mm Whatman paper, cover with Saran Wrap, and dry in a gel dryer (Bio-Rad). Quantitate the labeling of substrate proteins by autoradiography.

For the IRAK1 kinase assay, refer to the protocol for the IRAK4 kinase assay, but use anti-IRAK1 monoclonal antibody (Santa Cruz Biotechnology, Santa Cruz, CA) for immunoprecipitation and 1 µg recombinant His-tagged Pellino 2 as a substrate instead.

3.6   ANALYSIS OF IKK SIGNALOSOME ACTIVATION

Upon innate immune stimulation (especially in response to TLR2, TLR4, and NOD2 activation), the IKK scaffolding protein, NEMO, becomes rapidly ubiquitinated.[44,45] Unlike most cellular polyubiquitinations, which typically lead to proteosomal degradation, NEMO ubiquitination is K63 linked and is thought to couple TAK1 to the IKKs such that TAK1 can phosphorylate the activation loop of IKKs, resulting in their activation.[45] Therefore, NEMO ubiquitination provides another way to determine the amplitude and time course of TLR signaling.

3.6.1   DETECTION OF NEMO UBIQUITINATION

Protocol 3.6: NEMO Ubiquitination

1. Stimulate cells with the ligand of choice. The time course will depend on the ligand-receptor interaction. For instance, in macrophages, LPS stimulation of TLR4 is rapid (0–20 minutes), while Poly I:C stimulation of TLR3 is slower (30–180 minutes), depending on the speed of endocytosis.
2. Lyse cells in RIPA buffer (50 mM Tris [pH 7.5], 150 mM NaCl, 0.25% DOC, 1% NP-40, 1% SDS, 1mM EDTA, 1mM EGTA, 1 mM β-glycerophosphate, 1 mM PSMF, 1 mM sodium orthovanadate, 1 μg/mL leupeptin, 1 μg/mL pepstatin, 10 nM calyculin A, 5 mM iodoacetimide, and 5 mM N-ethylmaleimide). *Note*: omit DTT or β-mercaptoethanol from the buffer.
3. Add anti-NEMO antibody. In our hands, the mouse monoclonal clone C73-764 available commercially from BD Biosciences (San Jose, CA) gives strong results.
4. Incubate for at least 4 hours with rotation at 4 °C.
5. Add protein G Sepharose beads that have been previously blocked with 3% BSA.
6. Incubate for at least 1 hour with rotation at 4 °C.
7. Spin tubes at 3,000 rpm at 4 °C.
8. Discard supernatant. Wash beads three times with RIPA buffer containing 1 M NaCl. This is performed to rid the NEMO immunoprecipitates of other NEMO-bound ubiquitinated proteins. Alternatively, to remove NEMO-bound ubiquitinated proteins, the original lysate is boiled for 5 minutes in the presence of 1% SDS; the mix is cooled down, then diluted tenfold (to bring SDS concentration to 0.1%); and immunoprecipitation is performed as described above.
9. Before performing SDS-PAGE, the salt concentration of the immunoprecipitate needs to be decreased by washing the beads three times in RIPA buffer containing 150 mM NaCl.
10. Run two separate Western blots using an anti-NEMO rabbit antibody (Cell Signaling Technology, Danvers, MA) and a monoclonal anti-ubiquitin antibody (P4D1; available commercially from Santa Cruz Biotechnology.

NEMO typically runs as a doublet at approximately 50 kD. Ubiquitination adds approximately 10–12 kD per chain. Thus, there should be a time-dependent smear at approximately 70–150 kD on the blots.[44,45] NEMO ubiquitination appears to be a low-stoichiometry event,[45] so the predominant bands will remain at approximately 50 kD.

3.6.2 Detection of IKK Activation

TLR activation induces IKK kinase activity downstream of NEMO ubiquitination. IKK kinase activity can be assayed either directly via the immunoprecipitation of the IKK complex (NEMO + IKK) followed by the addition of exogenous substrate (IκBα) or indirectly via Western blotting for phosphorylated IκBα.

Protocol 3.7: Determination of IKK Activity

1. After stimulation for 0–90 minutes, lyse cells (10^6) in 1 ml lysis buffer (20 mM Tris [pH 7.5], 150 nM NaCl, 1 mM EDTA, 1 mM EGTA, 1% Triton X-100, 2.5 mM sodium pyrophosphate, 1 mM β-glycerophosphate, 1 mM sodium orthovanadate, 1 μg/ml leupeptin, 1 mM PMSF, and 10 nM calyculin A). After incubation for 10 minutes at 4 °C, clear the lysate via centrifugation at 13,000 rpm for 10 minutes at 4 °C.

2. Immunoprecipitate IKK kinase complex with anti-NEMO antibody (mouse monoclonal clone C73-764; BD Biosciences, San Jose, CA) for 2 hours with rotation. Add protein G Sepharose beads, and incubate with rotation at 4 °C for an additional 45 minutes.

3. Spin tubes at 4,000 rpm for 1 minute. Discard supernatant. Wash Sepharose beads three times with 1 ml lysis buffer, by resuspension and spinning at 4,000 rpm for 1 minute. Repeat three times.

4. It is necessary to remove the detergent (Triton X-100) and the divalent chelators prior to performing the kinase assay. For this reason, wash the immunoprecipitates three times with 1× kinase buffer (25 mM Tris-HCl [pH 7.4], 100 mM NaCl, 1 mM β-glycerophosphate, 18 mM $MgCl_2$, and 1 mM DTT). Resuspend the beads in 100 μl of kinase buffer.

5. Mix 20 μl of bead suspension with 2 μg of GST-IκBα fusion protein (Santa Cruz Biotechnology, Santa Cruz, CA). Start the reaction by adding 3 μl of a 10× ATP mixture containing nonradioactive ATP (1 mM) and [γ-^{32}P]-labeled ATP (10 μCi). Mix and incubate at 30 °C for 30 minutes in a total reaction volume of 30 μl.

6. Stop the reaction by the addition of 30 μl of SDS sample buffer followed by boiling for 5 minutes. Resolve proteins on a 12% SDS-PAGE gel: transfer proteins onto nitrocellulose membrane (as for a Western blot) and expose to X-ray film for 30 minutes to overnight (depending on activity levels). Determine kinase activity by quantifying ^{32}P incorporation into the GST-IκBα fusion protein band.

7. Western-blot the nitrocellulose membrane to control for equivalent immunoprecipitation of the IKK complex using rabbit polyclonal anti-IKKα, anti-IKKβ, and anti-NEMO antibodies (Cell Signaling Technology, Danvers, MA).

In response to IKK activation, IκBα is rapidly phosphorylated and degraded. Historically, the disappearance of IκBα has been an important marker of IKK activation. Unfortunately, assaying IκBα disappearance by Western blotting requires essentially an all-or-none response.[35] While this works well for TNFα stimulation, less potent activators of the IKK complex do not cause enough IκBα degradation to be in a linear, assayable range by Western blotting.[35,38,45] To overcome this limitation, antibodies directed against phospho-S32/S36-IκBα have been developed. Because one is now assaying for the appearance of the IKK-phosphorylated form of IκBα, rather than the disappearance of IκBα, it is easier to be in linear range, and less potent activators of IKK can now be assayed.[41,45] These experiments are relatively straightforward to perform with simple stimulation of cells followed by lysis and Western blotting.

Protocol 3.8: Determination of IκBα Phosphorylation

1. Stimulate cells of choice with a desired agonist for 0, 5, 15, 30, 60, and 90 minutes.
2. Wash cells with PBS containing 10 nM calyculin (Millipore), and lyse with lysis buffer (20 mM Tris-HCl [pH 7.5], 150 nM NaCl, 1 mM EDTA, 1 mM EGTA, 1% Triton X-100, 2.5 mM sodium pyrophosphate, 1 mM β-glycerophosphate, 1 mM sodium orthovanadate, 1 μg/mL leupeptin, 1 mM PMSF, and 10 μM calyculin A). Incubate on ice for 20 minutes, and then spin at 13,000 rpm for 10 minutes. Discard the pellet.
3. Determine protein concentrations in lysates by the BCA assay (Pierce Biotechnology, Rockford, IL), and run approximately 75–100 μg of total protein on a 12% SDS-PAGE gel.
4. Transfer proteins onto a 45 micron PVDF membrane (Millipore), and block with 5% milk in TBST. After 1 hour, wash the membrane twice with TBST. Add anti-phospho-IκBα antibody (Cell Signaling Technology, Danvers, MA) diluted 1:1,000 in 3% BSA-TBST. Incubate with rotation overnight at 4 °C.
5. Wash the blot four times with TBST, and incubate with anti-mouse secondary antibody conjugated to HRP (diluted at 1:10,000 in 5% milk-TBST; Chemicon, Temecula, CA) for 1 hour at room temperature. Wash four times with TBST for 5 minutes each wash. Visualize by standard chemiluminescent techniques.

At early time points, there should be a dose- and time-dependent increase in phospho-IκBα, which can be visualized as a strong band at approximately 38 kD.

This band may decrease at later time points as the phosphorylated form of IκBα is degraded.

REFERENCES

1. Akira S. TLR signaling. *Curr. Top. Microbiol. Immunol.* 311, 1, 2006.
2. Muzio M., and Mantovani A. Toll-like receptors (TLRs) signalling and expression pattern. *J. Endotoxin. Res.* 7, 297, 2001.
3. Takeda K, and Akira S. Toll-like receptors in innate immunity. *Int. Immunol.* 17, 1, 2005.
4. Seki E., Tsutsui H., Tsuji N. M., Hayashi N., Adachi K., Nakano H., Futatsugi-Yumi-kura S., Takeuchi O., Hoshino K., Akira S., Fujimoto J., and Nakanishi K. Critical roles of myeloid differentiation factor 88-dependent proinflammatory cytokine release in early phase clearance of *Listeria monocytogenes* in mice. *J. Immunol.* 169, 3863, 2002.
5. Ray A., Karmakar P., and Biswas T. Up-regulation of CD80-CD86 and IgA on mouse peritoneal B-1 cells by porin of *Shigella dysenteriae* is Toll-like receptors 2 and 6 dependent. *Mol. Immunol.* 41, 1167, 2004.
6. Pasare C., and Medzhitov R. Control of B-cell responses by Toll-like receptors. *Nature* 438, 364, 2005.
7. Blander J. M., and Medzhitov R. Toll-dependent selection of microbial antigens for presentation by dendritic cells. *Nature* 440, 808, 2006.
8. Faure E., Thomas L., Xu H., Medvedev A., Equils O., and Arditi M. Bacterial lipopolysaccharide and IFN-gamma induce Toll-like receptor 2 and Toll-like receptor 4 expression in human endothelial cells: role of NF kappa B activation. *J. Immunol.* 166, 2018, 2001.
9. Lee J., Mo J. H., Katakura K., Alkalay I., Rucker A. N., Liu Y. T., Lee H. K., Shen C., Cojocaru G., Shenouda S., Kagnoff M., Eckmann L., Ben-Neriah Y., and Raz E. Maintenance of colonic homeostasis by distinctive apical TLR9 signalling in intestinal epithelial cells. *Nat. Cell Biol.* 8, 1327, 2006.
10. Muzio M., Bosisio D., Polentarutti N., D'amico G., Stoppacciaro A., Mancinelli R., van't Veer C., Penton-Rol G., Ruco L. P., Allavena P., and Mantovani A. Differential expression and regulation of toll-like receptors (TLR) in human leukocytes: selective expression of TLR3 in dendritic cells. *J. Immunol.* 164, 5998, 2000.
11. Gewirtz A. T., Navas T. A., Lyons S., Godowski P. J., and Madara J. L. Bacterial flagellin activates basolaterally expressed TLR5 to induce epithelial proinflammatory gene expression. *J. Immunol.* 167, 1882, 2001.
12. Zhang D., Zhang G., Hayden M. S., Greenblatt M. B., Bussey C., Flavell R. A., and Ghosh S. A toll-like receptor that prevents infection by uropathogenic bacteria. *Science* 303, 1522, 2004.
13. Hornung V., Guenthner-Biller M., Bourquin C., Ablasser A., Schlee M., Uematsu S., Noronha A., Manoharan M., Akira S., de Fougerolles A., Endres S., and Hartmann G. Sequence-specific potent induction of IFN-alpha by short interfering RNA in plasmacytoid dendritic cells through TLR7. *Nat. Med.* 11, 263, 2005.
14. Kawai T., and Akira S. Innate immune recognition of viral infection. *Nat. Immunol.* 7, 131, 2005.
15. Yumoto H., Chou H. H., Takahashi Y., Davey M, Gibson F. C. III, and Genco C. A. Sensitization of human aortic endothelial cells to lipopolysaccharide via regulation of Toll-like receptor 4 by bacterial fimbria-dependent invasion. *Infect. Immun.* 73, 8050, 2005.

16. Sanghavi S. K., and Reinhart T. A. Increased expression of TLR3 in lymph nodes during simian immunodeficiency virus infection: implications for inflammation and immunodeficiency. *J. Immunol.* 175, 5314, 2005.

17. de Bouteiller O., Merck E., Hasan U. A., Hubac S., Benguigui B., Trinchieri G., Bates E. E., and Caux C. Recognition of double-stranded RNA by human toll-like receptor 3 and downstream receptor signaling requires multimerization and an acidic pH. *J. Biol. Chem.* 280, 38133, 2005.

18. Leifer C. A., Brooks J. C., Hoelzer K., Lopez J., Kennedy M. N., Mazzoni A., and Segal D. M. Cytoplasmic targeting motifs control localization of toll-like receptor 9. *J. Biol. Chem.* 281, 35585, 2006.

19. Barton G. M., Kagan J. C., and Medzhitov R. Intracellular localization of Toll-like receptor 9 prevents recognition of self DNA but facilitates access to viral DNA. *Nat. Immunol.* 7, 49, 2006.

20. Latz E., Schoenemeyer A., Visintin A., Fitzgerald K. A., Monks B. G., Knetter C. F., Lien E., Nilsen N. J., Espevik T., and Golenbock D. T. TLR9 signals after translocating from the ER to CpG DNA in the lysosome. *Nat. Immunol.* 5, 190, 2004.

21. Honda K., Ohba Y., Yanai H., Negishi H., Mizutani T., Takaoka A., Taya C., and Taniguchi T. Spatiotemporal regulation of MyD88-IRF-7 signalling for robust type-I interferon induction. *Nature* 434, 1035, 2005.

22. Li X., and Qin J. Modulation of Toll-interleukin 1 receptor mediated signaling. *J. Mol. Med.* 83, 258, 2005.

23. Qin J., Jiang Z., Qian Y., Casanova J. L., and Li X. IRAK4 kinase activity is redundant for interleukin-1 (IL-1) receptor-associated kinase phosphorylation and IL-1 responsiveness. *J. Biol. Chem.* 279, 26748, 2004.

24. Jiang Z., Ninomiya-Tsuji J., Qian Y., Matsumoto K., and Li X. Interleukin-1 (IL-1) receptor-associated kinase-dependent IL-1-induced signaling complexes phosphorylate TAK1 and TAB2 at the plasma membrane and activate TAK1 in the cytosol. *Mol. Cell Biol.* 22, 7158, 2002.

25. Qin J., Qian Y., Yao J., Grace C., and Li X. SIGIRR inhibits interleukin-1 receptor- and toll-like receptor 4-mediated signaling through different mechanisms. *J. Biol. Chem.* 280, 25233, 2005.

26. Jiang Z., Johnson H. J., Nie H., Qin J., Bird T. A., and Li X. Pellino 1 is required for interleukin-1 (IL-1)-mediated signaling through its interaction with the IL-1 receptor-associated kinase 4 (IRAK4)-IRAK-tumor necrosis factor receptor-associated factor 6 (TRAF6) complex. *J. Biol. Chem.* 278, 10952, 2003.

27. Sebban H., Yamaoka S., and Courtois G. Posttranslational modifications of NEMO and its partners in NF-kappaB signaling. *Trends Cell Biol.* 16, 569, 2006.

28. Yao J., Kim T. W., Qin J., Jiang Z., Qian Y., Xiao H., Lu Y., Qian W., Gulen M. F., Sizemore N., DiDonato J., Sato S., Akira S., Su B., and Li X. Interleukin-1 (IL-1)-induced TAK1-dependent versus MEKK3-dependent NFkappaB activation pathways bifurcate at IL-1 receptor-associated kinase modification. *J. Biol. Chem.* 282, 6075, 2007.

29. Qin J., Yao J., Cui G., Xiao H., Kim T. W., Fraczek J., Wightman P., Sato S., Akira S., Puel A., Casanova J. L., Su B., and Li X. TLR8-mediated NF-kappaB and JNK activation are TAK1-independent and MEKK3-dependent. *J. Biol. Chem.* 281, 21013, 2006.

30. Whitehead Institute for Biomedical Research. Primer3: WWW primer tool. http://biotools.umassmed.edu/bioapps/primer3_www.cgi (accessed January 16, 2008).

31. Seibl R., Birchler T., Loeliger S., Hossle J. P., Gay R. E., Saurenmann T., Michel B. A., Seger R. A., Gay S., and Lauener R. P. Expression and regulation of Toll-like receptor 2 in rheumatoid arthritis synovium. *Am. J. Pathol.* 162, 1221, 2003.

32. Shiratsuchi A., Watanabe I., Takeuchi O., Akira S., and Nakanishi Y. Inhibitory effect of Toll-like receptor 4 on fusion between phagosomes and endosomes/lysosomes in macrophages. *J. Immunol.* 172, 2039, 2004.

33. Uematsu S., Sato S., Yamamoto M., Hirotani T., Kato H., Takeshita F., Matsuda M., Coban C., Ishii K. J., Kawai T., Takeuchi O., and Akira S. Interleukin-1 receptor-associated kinase-1 plays an essential role for Toll-like receptor (TLR) 7- and TLR9-mediated interferon-{alpha} induction. *J. Exp. Med.* 201, 915, 2005.

34. Qian Y., Commane M., Ninomiya-Tsuji J., Matsumoto K., and Li X. IRAK-mediated translocation of TRAF6 and TAB2 in the interleukin-1-induced activation of NFkappa B. *J. Biol. Chem.* 276, 41661, 2001.

35. Sato S., Sanjo H., Takeda K., Ninomiya-Tsuji J., Yamamoto M., Kawai T., Matsumoto K., Takeuchi O., and Akira S. Essential function for the kinase TAK1 in innate and adaptive immune responses. *Nat. Immunol.* 6, 1087, 2005.

36. Jacinto R., Hartung T., McCall C., and Li L. Lipopolysaccharide- and lipoteichoic acid-induced tolerance and cross-tolerance: distinct alterations in IL-1 receptor-associated kinase. *J. Immunol.* 168, 6136, 2002.

37. Williams K. L., Lich J. D., Duncan J. A. Reed W., Rallabhandi P., Moore C., Kurtz S., Coffield V. M., Accavitti-Loper M. A., Su L., Vogel S. N., Braunstein M., and Ting J. P. The CATERPILLER protein monarch-1 is an antagonist of Toll-like receptor-, tumor necrosis factor alpha-, and Mycobacterium tuberculosis-induced pro-inflammatory signals. *J. Biol. Chem.* 280, 39914, 2005.

38. Yamamoto M., Sato S., Hemmi H., Uematsu S., Hoshino K., Kaisho T., Takeuchi O., Takeda K., and Akira S. Essential role for TIRAP in activation of the signalling cascade shared by TLR2 and TLR4. *Nature* 420, 324, 2002.

39. Yang K., Puel A., Zhang S., Eidenschenk C., Ku C. L., Casrouge A., Picard C., von Bernuth H., Senechal B., Plancoulaine S., Al-Hajjar S., Al-Ghonaium A., Marodi L., Davidson D., Speert D., Roifman C., Garty B. Z., Ozinsky A., Barrat F. J., Coffman R. L., Miller R. L., Li X., Lebon P., Rodriguez-Gallego C., Chapel H., Geissmann F., Jouanguy E., and Casanova J. L. Human TLR-7-, -8-, and -9-mediated induction of IFN-alpha/beta and -lambda Is IRAK-4 dependent and redundant for protective immunity to viruses. *Immunity* 23, 465, 2005.

40. Lye E., Mirtsos C., Suzuki N., Suzuki S., and Yeh W. C. The role of interleukin 1 receptor-associated kinase-4 (IRAK-4) kinase activity in IRAK-4-mediated signaling. *J. Biol. Chem.* 279, 40653, 2004.

41. Kim T. W., Staschke K., Bulek K., Yao J., Peters K., Oh K. H., Vandenburg Y., Xiao H., Qian W., Hamilton T., Min B., Sen G., Gilmour R., and Li X. A critical role for IRAK4 kinase activity in Toll-like receptor-mediated innate immunity. *J. Exp. Med.* 204, 1025, 2007.

42. Sun L., Deng L., Ea C. K., Xia Z. P., and Chen Z. J. The TRAF6 ubiquitin ligase and TAK1 kinase mediate IKK activation by BCL10 and MALT1 in T lymphocytes. *Mol. Cell* 14, 289, 2004.

43. Strelow A., Kollewe C., and Wesche H. Characterization of Pellino2, a substrate of IRAK1 and IRAK4. *FEBS Lett.* 547, 157, 2003.

44. Abbott D. W., Wilkins A., Asara J. M., and Cantley L. C. The Crohn's disease protein, NOD2, requires RIP2 in order to induce ubiquitinylation of a novel site on NEMO. *Curr. Biol.* 14, 2217, 2004.

45. Abbott D. W., Yang Y., Hutti J. E., Madhavarapu S., Kelliher M. A., and Cantley L. C. Coordinate regulation of Toll-like Receptor and NOD2 signaling by K63-linked polyubiquitin chains. *Mol. Cell Biol.* 27, 6012, 2007.

4 Intracellular Trafficking of Toll-Like Receptors

Harald Husebye, Øyvind Halaas,
Harald Stenmark, and Terje Espevik

CONTENTS

4.1 INTRODUCTION

4.1.1 TOLL-LIKE RECEPTORS

The immediate response to an invading organism is coordinated by the innate immune system. The pathogens are sensed by a family of germ line–encoded receptors recognizing conserved molecular patterns shared by large groups of pathogens. These types of ligands are often referred to as pathogen-associated molecular patterns (PAMPs) and are mainly derived from the microbial nucleic acids or cell walls. The PAMPs are recognized by members in the family of Toll-like receptors (TLRs), expressed in monocytes, macrophages, and dendritic cells. Functional TLRs are essential for our defense against microbes and initiate the inflammatory reactions. Uncontrolled TLR signaling may lead to life-threatening complications such as systemic inflammatory response syndrome (SIRS; also often referred to as septic shock), chronic inflammation, and autoimmunity. The TLR family consists of at least 10 members in humans, where they sense microbial compounds such as lipopolysaccharide (TLR4) and lipoproteins (TLR2, which dimerizes with TLR1 or TLR6), proteins (flagellin; TLR5), and viral and bacterial nucleic acids (TLR3, TLR7, TLR8, and TLR9).[1,2]

The different TLRs show distinct spatial distribution within the cell and, as a consequence, different trafficking patterns. TLR2 and TLR4 are expressed on the plasma membrane, on endosomes, and as a large pool in the Golgi area. TLRs 3, 7, 8, and 9 reside in the endoplasmic reticulum (ER) in unstimulated cells. In contrast to TLR2 and TLR4, whose ligand binding and initiation of signaling occur at the plasma membrane, TLRs 3, 7, 8, and 9 are only intracellular and are mobilized from the ER to endosomes, where they bind their cognate ligands.

The LPS receptor complex consists of TLR4 and the co-receptors MD-2 and CD14. MD-2 is a small secreted glycoprotein that is essential for TLR4 signaling. CD14 appears both in a secreted form and attached to cellular membranes via a glycosylphosphatidylinositol (GPI) anchor. CD14 enhances LPS signaling, probably by providing a more efficient delivery of LPS to TLR4/MD2. LPS initiates a signaling cascade through the TIR adaptors MyD88 (myeloid differentiation factor 88), TIRAP (TIR domain–containing adaptor protein; also known as Mal), TRIF (Toll/IL-1R [TIR] domain–containing adaptor inducing IFN-β), and TRAM (TRIF-related adaptor molecule), eventually leading to nuclear translocation of the proinflammatory transcription factor NF-κB.[3] Excessive host responses toward *Meningococcal* bacteria may lead to life-threatening complications such as septic shock, multiorgan failure, and death.[4] Thus, LPS-induced host responses must be tightly controlled and time limited, and are achieved by regulating TLR and signaling component levels and availability. Desensitization to LPS in human monocytes is caused partly by downregulation of surface TLR4/MD-2.[5]

4.1.2 SORTING SIGNALS

The lysosomal targeting of transmembrane proteins is determined by tyrosine phosphorylation, di-leucine motifs, and ubiquitination.[6] Tyrosine-based YXXØ motifs can be found in TLR4 (587YDAF, 622YRDF, and 707YLEW), as can a di-leucine motif [DE]XXXL[LI] (697ELYRLL). These motifs are recognized by clathrin and endocytosis adaptor proteins (AP) in receptor endocytosis. The covalent attachment of ubiquitin to a lysine residue in a substrate protein proceeds through sequential action of ubiquitin-activating (E1), -conjugating (E2), and -ligating (E3) enzymes.[7] Ubiquitination occurs following the tyrosine phosphorylation of the receptor. It has been shown that a single ubiquitin is sufficient for directing epidermal growth factor receptors (EGFRs) for degradation.[8] However, recent published work shows that Lys63-linked polyubiquitin chains also serve a role as sorting signals for lysosomal degradation.[9] Ubiquitinated receptors are recognized by a family of proteins carrying the ubiquitin-interacting motif (UIM) involved in the control of internalization and the lysosomal targeting of ubiquitinated proteins. The hepatocyte growth factor–regulated tyrosine kinase substrate (Hrs) plays a key role in recognizing and targeting ubiquitinated receptors for lysosomal degradation and is a member of the UIM protein family.[10] Hrs carries the phosphatidylinositol-3-phosphate (PtdIns[3]P-) binding domain FYVE (i.e., conserved in *Fab*1, *YOTB*, *Vac*1, and *EEA*1).[11] PtdIns(3)P is highly enriched on early sorting endosomes and is generated by the action of the class III PI3-kinase, hVPS34. The clathrin box motif of Hrs allows it to assemble in clathrin-rich microdomains of the early sorting endosomal-limiting membrane.[12] We have recently shown that Hrs is essential for targeting TLR4 for lysosomal degradation resulting in the termination of signaling.[13] Tollip is a negative regulator of TLR and IL-1 receptor signaling, and it has been shown that Tollip sorts ubiquitinated

IL-1R toward lysosomal degradation.[14] The function of Tollip in lysosomal degradation of the TLRs has yet to be solved.

4.1.3 STABLE CELL LINES

As mentioned above, the TLRs are expressed on innate immune cells such as monocytes, macrophages, and dendritic cells. Although relatively easy to obtain from human peripheral blood, these cells are hard to genetically manipulate. Furthermore, good antibodies toward many of the TLRs have proven difficult to obtain as all receptors contain conserved leucine-rich repeats and are heavily glycosylated. As a consequence, stable HEK293 cells' transfectants expressing fusions between the TLRs and fluorescent proteins provide valuable tools for the intracellular trafficking of TLRs. Furthermore, as long as the fluorescent proteins are fused to the C-termini of the TLRs, the TLRs are functional, allowing studies of both trafficking and signaling events in the HEK293 cells.[1,13,15]

4.2   CELLULAR TRAFFICKING

At the plasma membrane, there is constant endocytosis of membrane and nonactivated transmembrane proteins; whereas activated ligand-bound receptors are targeted for destruction, the nonactivated may recycle back to the plasma membrane. Confocal microscopy at high speed and resolution using the LSM 5 LIVE laser-scanning unit shows a rapid trafficking of TLR4 between the plasma membrane and endosomes (Husebye and Espevik, unpublished). Signaling speeds up endocytosis and enlarges the TLR4 positive endosomes.[13] The absolute requirement for MD 2 in TLR4 signaling allows the study of endocytosis and intracellular trafficking of TLR4 under signaling conditions and conditions where signaling cannot occur (i.e., due to lack of MD2). CD14 greatly increases the binding to the surface and uptake of LPS, as has previously been reported.[5,15] As the HEK293 cells do not express endogenous CD14, the CD14 transfectants are easily identified by a rapid accumulation of fluorescently labeled LPS at the plasma membrane upon stimulation (Figure 4.1).

Invagination of membranes is caused by recruitment of clathrin to microdomains through the clathrin adaptors, pulling the membrane into the cell. A frequently used marker for studying receptor-mediated and clathrin-dependent endocytosis is fluorescently labeled transferrin. Transferrin is endocytosed by binding to the transferrin receptor (Tf-R). The endocytosis of transferrin bound to the Tf-R requires the formation of clathrin-coated pits and the action of dynamin.[16] This allows the use of fluorescent transferrin for identifying cells with knockdown of clathrin or dynamin since transferrin accumulates on the plasma membrane of these cells. Targeting the clathrin-heavy chain using siRNA technology is a powerful way of studying the involvement of clathrin in receptor endocytosis.[17] The transfectants where the clathrin-heavy chain has been knocked down show impaired endocytosis of the TLR4/LPS receptor complex.[13] Furthermore, the transfectants also showed increased NF-κB activation compared to cells treated with control siRNA, consistent with impaired endocytosis of the LPS receptor complex (Husebye and Halaas, unpublished data). In these types of experiments, care should be taken to avoid

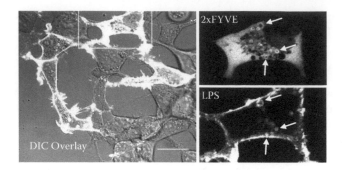

FIGURE 4.1   LPS Is Internalized on Early Sorting Endosomes

Note: HEK293 cells expressing TLR4[YFP]/MD-2 were co-transfected with CD14 and a marker for early sorting endosomes, 2xFYVE-CFP. Confocal images of cells 60 minutes after addition of Cy5-labeled LPS (250 ng/ml). Arrows point at early sorting endosomes positive for Cy5-labeled LPS. Bar = 20 μm.

overestimation of the cells being defective in endocytosis. As clathrin-dependent endocytosis is inhibited during mitosis,[18] a fluorescent nuclear stain should be used to exclude cells undergoing mitosis.

The dynamin GTPase is involved in the budding of the clathrin-coated endocytotic vesicles from the plasma membrane. Dynamin exists as two isoforms in mammalian cells, dynamin-I and dynamin-II. The dominant negative versions of dynamin, Dyn1K44A and Dyn2K44A, lack the ability of binding GTP and are both potent inhibitors of receptor-mediated endocytosis.[19] The dominant negative versions of dynamin have frequently been used to define receptor-mediated uptake.

The nonfunctional ubiquitin UbR lacks the two C-terminal glycines required for conjugation of Ub to lysine residues and blocks EGFR endocytosis upon co-transfection.[20] To investigate the effect of ubiquitination of TLR4 on the endocytosis of the LPS receptor complex cells, we transfected cells with UbR and compared the effect to cells overexpressing the Dyn1K44A mutant (Figure 4.2). For the detection of Dyn1K44A transfectants showing defective transferrin uptake using confocal microscopy, a period of 5–6 days of transfection was required, although 3 days were sufficient to affect the activation of NF-κB using the Dyn1K44A and UbR mutants.[13]

Protocol 4.1: Plasmid DNA Transfection

1. Maintain the cells in DMEM supplemented with 10% FCS, 0.5 mg/ml G418, and 10 μg/ml Ciprofloxacin, and do not allow the cells to grow to a confluency of more than 50% before splitting.
2. One the day before transfection, seed 250,000–300,000 cells per 35 mm dish (for imaging, use the glass-bottom 35mm γ-irradiated cell culture dishes from MatTek, Ashland, MA).
3. Transfect the cells using 3 μl GeneJuice (Novagen, Darmstadt, Germany) per 1 μg DNA in 100 μl Opti-MEM I Reduced Serum Medium (Gibco, Carls-

bad, CA), according to the manufacturer's protocol. The expression vectors on which the respective transient transfection protocols are developed have been described before.[13] When working with the LPS receptor complex, it is important to avoid endotoxin contamination in the plasmid DNA preparations; to avoid this, use the endotoxin-free plasmid DNA isolation kits (Qiagen, Valencia, CA).

4. For the introduction of the 2xFYVE-CFP (cyan fluorescent protein) probe, do not allow the amount to exceed 50% of the total transfected plasmid DNA per dish.

Protocol 4.2: Confocal Microscopy

1. Image live cells at 37 °C. We use an Axiovert 200-M microscope with a heated stage equipped with a Zeiss LSM 510 META scanning unit and an LSM 5 LIVE laser-scanning unit.
2. Objectives used are the Zeiss 63X plan apochromat or a 100X oil alpha Plan-Fluar with numeric apertures of 1.4 and 1.45, respectively.
3. Image fluorescence in multiple channels using individual tracks if cross-bleeding between channels is observed. Using the Zeiss LSM 510 META scanning unit, Hoechst 33342 is excited at 405 nm with emission at 475–525 nm, CFP is excited at 458 nm with emission at 470–500 nm, yellow fluorescent protein (YFP) is excited at 514 nm with emission at 535–590 nm, LysoTracker Red is excited at 543 nm with emission at 565–615 nm, Cy5 is excited at 633 nm with emission above 650 nm, and Draq5 is excited at 633 nm with emission above 650 nm.
4. Using the LSM 5 LIVE laser-scanning unit, CFP is excited at 405 nm with emission at 415–480 nm, YFP is excited at 489 nm with emission above 505 nm, and Cy5 is excited at 635 nm with emission above 650 nm.

Protocol 4.3: Labeling LPS and Antibodies with Fluorescent Dyes

1. Label 5 mg ultrapure *Escherichia coli* LPS (0111:B4 LPS, Invivogen, San Diego, CA) with Cy5 by using the Cy5-mono reactive dye kit (Amersham Biosciences, Piscataway, NJ), according to the manufacturer's protocol.
2. Let the reaction proceed for 1 hour at room temperature, and remove the unincorporated dye from the labeled LPS by running the reaction mix through a 10 ml PD10 gel filtration column (Amersham Biosciences).
3. Collect the Cy5-labeled LPS in aliquots of 100 µl in PBS, and determine the concentration by using the LAL kit from BioWhittaker (Walkersville, MD), according to the manufacturer's instructions.
4. Label monoclonal antibodies with either the Alexa Fluor protein-labeling kit or the Alexa Fluor microscale protein-labeling kit from Invitrogen (Carls-

bad, CA), according to the manufacturer's protocol (use the microscale kit if a small amount of antibody is to be labeled or if the concentration of the antibody solution is below 0.5 mg/ml). Unlabeled dye can be removed by running the reaction mix on a gel filtration system (e.g., a 1 ml desalting column) or by using the supplied columns. We use the SMART system from Amersham Biosciences.

5. Fluorescently labeled ligands for other TLRs, such as TLR9 phosphothioate CpG DNA labeled with OregonGreen or TexasRed, can be purchased from MWG Biotech (Ebersberg, Germany). Labeled double-stranded RNA can be ordered from Dharmacon (Lafayette, CO).

Protocol 4.4: Preparation of Cells for Live Cell Imaging

1. Remove the transfection mix from the transfected cells, and wash twice with 1 ml DMEM without phenol red (BioWhittaker, Walkersville, MD) supplemented with 10% FCS (Euroclone, Milan, Italy) and 25 mM HEPES (Gibco, Carlsbad, CA).

2. Sonicate the Cy5-labeled LPS in a sonicator water bath for 1 minute. Add 100–250 ng Cy5-labeled LPS to each 35 mm dish containing 1 ml medium. For the study of LPS and transferrin co-trafficking, Alexa Fluor 546-labeled transferrin (5 μg/ml) can be added simultaneously with Cy5-labeled LPS.

3. For the identification of acidic compartments (pH < 5.0) such as the late endosomes, LysoTracker Red or LysoTracker Green (Invitrogen, Carlsbad, CA) can be used. At the desired time point following stimulation, add 25 nM LysoTracker in 1 ml medium. Incubate for 30 minutes, and wash the cells two times in DMEM without phenolred supplemented with 10% FCS and 25 mM HEPES.

4. Visualize the nucleus in living cells using 1–2 μM Hoechst 33342 (Invitrogen, Carlsbad, CA) or 0.5–1 μM Draq5 (Biostatus, Leicestershire, UK).

5. The endocytosis of LPS can be studied in live monocytes by adding Cy5-labeled LPS (100–250 ng), and co-trafficking of LPS and TLR4 during endocytosis can be studied by simultaneously adding Cy5-labeled LPS and an Alexa-labeled antibody directed toward TLR4 (HTA125, eBioscience, San Diego, CA). Add Cy5-labeled LPS (100–250 ng/ml) and Alexa-labeled HTA125 (5 μg/ml). To minimize the amount of antibody used, apply 80 μl of antibody solution at the center of the 35 mm dish. Include an Alexa-labeled IgG2a (eBioscience) as a control for nonspecific binding.

FIGURE 4.2 Endocytosis of LPS Requires Dynamin but Not Ubiquitin

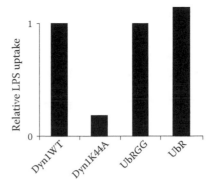

Note: Endocytosis of the LPS receptor complex is dependent on dynamin and clathrin. HEK293 cells expressing TLR4^{YFP}/MD-2 were transfected with 0.4 µg of CD14 together with 0.6 µg of c-myc-tagged dynamin WT, Dyn1K44A, UbRGG, or UbR and left for 6 days before analysis. The cells were simultaneously given Cy5-labeled LPS (250 ng/ml) and Alexa Fluor 546-labeled transferrin (5 µg/ml), and were fixed in 3% paraformaldehyde after 75 minutes of stimulation. The LPS uptake in Dyn1K44A ($n = 52$), UbRGG ($n = 54$), and UbR ($n = 44$) was compared to that of dynamin WT ($n = 46$) transfected cells and presented as relative LPS uptake. The LPS uptake was quantified using the LSM510 software (Zeiss, Göttingen, Germany) by measuring total intracellular fluorescence compared to the amount of LPS in the plasma membrane in manually drawn regions of interest.

Protocol 4.5: Preparation and Staining of Fixed Cells

1. Prepare the cells as described in Protocol 4.4.
2. Add 15% paraformaldehyde directly into the medium to a final concentration of 3%, and incubate for 30 minutes at room temperature.
3. Permeabilize the cells for 10 minutes using PEM buffer (80 mM K-Pipes [pH 6.8], 5 mM EGTA, 1 mM MgCl$_2$, and 0.05% saponin) on ice.
4. Block the unspecific binding of antibodies in blocking buffer (50 mM NH$_4$Cl, 1% BSA, and 0.05% saponin in PBS) at room temperature.
5. Dilute the actual antibodies in blocking buffer to a concentration of 2–5 µg/ml, and add 80 µl of the solution to the center of the dish. Incubate for 30–60 minutes.
6. Wash three times with 1 ml of washing buffer (0.05% saponin in PBS) before imaging. When unlabeled antibodies are used, wash two times in 1 ml PBS and then once in 1 ml blocking buffer. Add the fluorescently labeled secondary antibody (2 µg/ml), and incubate for 15 minutes. Wash three times in 1 ml of washing buffer.
7. For visualizing the nucleus, add 1 µM Draq5 (Biostatus, Leicestershire, UK).
8. For the staining of adhered human monocytes, the same protocol can be used with some adjustments of the protocol. Wash the adhered monocytes three times in 1 ml of Hank's buffered salt solution, and fix the cells in 3% paraformaldehyde dissolved in PBS. Include 2–5% A+ in the blocking buffer, and stain the cells as described above. We obtain endotoxin-free A+ serum for culturing human primary cells from the local blood bank.[13]

4.3   IDENTIFICATION OF EARLY SORTING ENDOSOMES

The early sorting endosomes are highly enriched in PtdIns(3)P that attracts proteins carrying the PtdIns(3)P-binding domain FYVE. While single FYVE domains derived from EEA1 or Hrs are mainly cytosolic, a tandem FYVE domain from Hrs (2×FYVE) has been successfully used for the identification of early sorting endosomes,[21] and 2×FYVE-GFP (green fluorescent protein) gives a very bright signal. For the identification of early endosomes in TLR4-YFP-expressing cells, we had to substitute enhanced green fluorescent protein (EGFP) with enhanced cyan fluorescent protein (ECFP) to better separate it spectrally from YFP. The internalization of Cy5-labeled LPS on 2×FYVE positive early sorting endosomes is shown in Figure 4.1 (right panels). At high expression levels, 2×FYVE transfectants show a significantly delayed endocytic trafficking of the LPS receptor complex (that is, when the coappearance of TLR4 and Cy5-labeled LPS on endosomes is investigated).

We have also observed that TLR4 accumulates on the 2×FYVE-containing endosomes in the cells showing delayed endocytic trafficking.[13] This is probably due to sequestration of PtdIns(3)P by 2×FYVE and has been described before.[22] To avoid this problem, the amount of transfected 2×FYVE-CFP can be reduced to 15–50% of the total plasmid DNA transfected to each 35 mm dish, or one may use the EEA1-CFP[23] instead. The EEA1-CFP does not seem to affect the kinetics of LPS endocytosis.

4.4   ENDOSOMAL SORTING AND LYSOSOMAL DEGRADATION

Hrs locates to early sorting endosomes, where it recognizes ubiquitinated receptors and targets them for lysosomal degradation in concert with the endosomal sorting complex required for transport (ESCRT) I–III complex. Figure 4.3 shows that coexpression of the dominant negative Hrs S270E,[24] or knockdown of Hrs by siRNA technology, results in increased NF-κB activation compared to control cells (Figure 4.3). A similar increase in LPS-induced signaling is observed in cells defective in endocytosis or overexpression of mutant ubiquitin UbR.[13] As Hrs or ubiquitination of TLR4 does not seem to affect the endocytosis of the LPS receptor complex, this is most probably due to defective endosomal sorting, leading to impaired lysosomal targeting of the activated LPS receptor complex by Hrs. In contrast to the Dyn1K44A and UbR mutants, the Hrs S270E mutant required 24 hours of transfection to affect the LPS-induced NF-κB activation. Prolonged periods of transfection gave a high constitutive background of NF-κB activation, masking the LPS-induced effect. This effect is observed in both Hrs wild-type and Hrs S270E transfected cells. Furthermore, it is important to keep c-myc-Hrs overexpression at a moderate level for studying the function of Hrs in receptor degradation. This applies to the studies of both EGFR[25] and TLR4[13] degradation. High levels of c-myc-Hrs expression seem to inhibit rather than promote degradation of ubiquitinated receptors.[22,24]

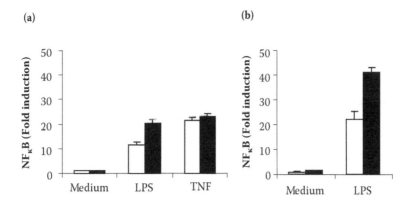

FIGURE 4.3 Inhibition of Hrs Results in Increased Activation of NF-κB

Note: (A) HEK293 cells expressing TLR4^{YFP}/MD-2 were co-transfected with CD14 and Hrs wild type or dominant negative Hrs S270E for 24 hours. The cells were stimulated with 10 ng/ml of LPS and assayed for NF-κB-induced luciferase expression after 5 hours. (B) HEK293 cells expressing TLR4^{YFP}/MD-2 were treated with 40 nM of siRNA targeting Hrs or with control siRNA for 48 hours before being seeded in 96-well plates. The cells were allowed to rest for 24 hours before being co-transfected with CD14 and Elam-luc for 24 hours. The cells were stimulated with 10 ng/ml of LPS for 5 hours before the analysis of NF-κB-induced luciferase expression.

Protocol 4.6: Plasmid DNA Transfection and NF-κB Activation

1. Seed 15,000–20,000 cells per well in 96 well plates. Use 100 µl DMEM medium with 10% FCS and 0.5 mg/ml G418 per well.
2. Dilute the plasmid DNA to 0.1 µg/µl in sterile Millipore H$_2$O prior to transfection, and transfect the cells using GeneJuice, according to the manufacturer's protocol (we use 0.3 µl GeneJuice and 0.1 µg DNA in 10 µl Opti-MEM [Gibco, Carlsbad, CA] per well, including 50 ng of ELAM-luc).
3. If two or more plasmids are to be co-transfected, lower the amount of the NF-κB reporter. Add similar amounts of the other plasmids to be transfected so that the total DNA amount adds up to 100 ng plasmid DNA per well. Use "empty" plasmid DNA with the same vector backbone as the effector plasmid if it is necessary to adjust the DNA concentration (up to 100 ng per well). If the effector is encoded as a "GFP" variant chimera, use the same vector backbone carrying the identical GFP variant to adjust the DNA concentration.
4. Before stimulation of the cells, wash the cells twice in DMEM supplemented with 10% FCS (from Euroclone, Milan, Italy) without antibiotics, and adjust the volume to 90 µl per well.
5. Sonicate LPS in a water bath sonicator for 1 minute, and dilute LPS in a medium containing 10% FCS to a concentration that is 10 times the concentration to be used for stimulation (we normally stimulate with 10–100 ng/ml LPS for 5–6 hours). Preincubate the LPS solution at 37 °C for 5 minutes.

6. Add 10 µl of the LPS solution to each well, and incubate the cells (using 10 ng of LPS, we observed that the activation of NF-κB takes at least 2 hours to exceed the level of the unstimulated control, in HEK293 cells expressing TLR4[YFP], MD-2, and CD14).[13]

7. After stimulation for the desired time period, remove the medium and add 50 µl cell lysis reagent (Promega, Madison, WI) to each well. Incubate for 5 minutes, and freeze at −20 °C (this gives a more efficient cell lysis).

8. On the day of NF-κB analysis, thaw cells for at least 1 hour at room temperature, and transfer 10 µl from each sample into a nontransparent white 96-well plate (Optiplate-96, PerkinElmer, Waltham, MA). (The use of a nontransparent plate reduces the cross talk between the wells during the monitoring of luciferase activity.)

9. Add 50 µl luciferase substrate equilibrated to room temperature to each well. The luciferase assay kit can be purchased from Promega. Immediately analyze the samples in a plate reader using the luminescence mode (we use the Viktor3 plate reader from PerkinElmer and record bioluminescence for a period of 1 second per well).

Protocol 4.7: Transfection of siRNA

In our hands, simultaneous transfection of siRNA and plasmid DNA into HEK293 cells leads to high mortality of the cells. We therefore first transfect the HEK293 cells with siRNA, and then plate them for plasmid DNA transfection. The siRNA duplexes are available from a number of sources; we used Dharmacon (Lafayette, CO) for custom siRNA synthesis and Qiagen (Valencia, CA) for the high-performance validated siRNAs.

1. Seed $0.5–1.0 \times 10^6$ cells in 25 cm^2 culture bottles. Allow the cells to grow for 24 hours. To transfect HEK293 with siRNA, use Oligofectamine (Invitrogen, Carlsbad, CA). The original protocol[26,27] is adapted for optimal transfection and performance of HEK293 cells.

2. Before transfection, wash the cells twice in DMEM with 0.5% FCS, and add 2.1 ml of DMEM with 0.5% FCS to the cells. Dilute the siRNA to the desired concentration in 455 µl Opti-MEM (Gibco, Carlsbad, CA).

3. Combine 7.8 µl Oligofectamine with 31.2 µl Opti-MEM per culture bottle to be transfected, and incubate for 10 minutes at room temperature.

4. Add the Oligofectamine mix to the siRNA, and allow complex formation for 15–20 minutes. Add the transfection mix to the cells, and incubate for 48 hours.

5. Split the cells, and seed in the appropriate well format. After 24 hours, transfect plasmid DNA using GeneJuice.

Protocol 4.8: Western Blotting and Immunoprecipitation

1. Harvest the cells by lysis at a confluency of 70–90%.
2. Add 300–400 µl lysis buffer (20–50 mM Tris-HCl [pH 8.0], 1 mM EDTA [pH 8.0], 1 mM EGTA [pH 8.0]), 137 mM NaCl, 1% Triton X-100, 1 mM sodium deoxycholate, 10% glycerol, 1 mM Na_3VO_4, 50 mM NaF, and Complete protease inhibitor [Roche]) per 35 mm well.
3. Incubate for 30 minutes on ice, and centrifuge at 13,000 rpm for 15 minutes at 4 °C. Transfer the supernatant to a new tube.
4. For Western blotting, denature the lysates in 1×NuPAGE LDS sample buffer supplemented with 25 mM DTT for 10 minutes at 70 °C.
5. For the harvesting of immune complexes, preclear the lysates by adding 25 µl protein G Sepharose (50% slurry), and incubate for 1 hour on a rotator.
6. Remove the beads by a 30-second centrifugation, combine the supernatant with the appropriate antibody (2–10 µg/ml), and incubate on a rotator for 1–2 hours. For this immunoprecipitation protocol, we have successfully used rabbit polyclonal antibody toward full-length GFP (Clontech, Mountain View, CA) and Hrs,[24] and the polyclonal rabbit antibody directed toward the N-terminal of TLR4 (eBioscience, San Diego, CA).
7. Harvest the immune complexes by adding 50 µl protein G Sepharose (50% slurry) for 2 hours or more, and wash four times in 0.5× lysis buffer.
8. Elute the immune complexes from the beads by adding 50–100 µl 2×LDS sample buffer (containing 1× lysis buffer supplemented with 50 mM DTT), and incubate at 95 °C for 5 minutes.
9. In cases where unspecific binding is observed, block the protein G Sepharose beads with BSA. Prepare the beads by an overnight incubation with 10 mg/ml ELISA-grade BSA on a rotator. In this case, skip the initial preclearing. Run the samples on PAGE, blot the samples, and develop the filters using ECL (Amersham Biosciences, Piscataway, NJ).
10. For the detection of TLRs, we run the samples on 7% Tris-acetate NuPAGE Novex polyacrylamide gels from Invitrogen (Carlsbad, CA).
11. Blot the samples onto 0.45 mm nitrocellulose filters from BioRad laboratories according the manufacturer's protocol.

4.5   LIST OF SUPPLIERS

Amersham Biosciences: http://www.gelifesciences.com
Axis-Shield: http://www.axis-shield.com
BioRad Laboratories: http://www.bio-rad.com
Biostatus: http://www.biostatus.com
BioWhittaker: http://www.lonza.com
Clontech: http://www.clontech.com
Dharmacon: http://www.dharmacon.com
eBioscience: http://www.ebioscience.com

Euroclone: http://www.euroclone.net
Gibco: http://www.invitrogen.com
Invitrogen: http://www.invitrogen.com
Invivogen: http://www.invivogen.com
MatTek: http://www.mattek.com
Molecular Probes: http://probes.invitrogen.com
MWG Biotech: http://www.mwg-biotech.com
Novagen: http://www.merckbiosciences.co.uk
PerkinElmer: http://www.perkinelmer.com
Promega: http://www.promega.com
Qiagen: http://www.qiagen.com
Roche: http://www.roche-applied-science.com

REFERENCES

1. Latz E., Schoenemeyer A., Visintin A., Fitzgerald K. A., Monks B. G., Knetter C. F., Lien E., Nilsen N. J., Espevik T., and Golenbock D. T. TLR9 signals after translocating from the ER to CpG DNA in the lysosome. Nat. Immunol. 5, 190, 2004.
2. Johnsen I. B., Nguyen T. T., Ringdal M., Tryggestad A. M., Bakke O., Lien E., Espevik T., and Anthonsen M. W. Toll-like receptor 3 associates with c-Src tyrosine kinase on endosomes to initiate antiviral signaling. EMBO J. 25, 3335, 2004.
3. Akira S., and Takeda K. Toll-like receptor signalling. Nat. Rev. Immunol. 4, 499, 2004.
4. Waage A., Brandtzaeg P., Halstensen A., Kierulf P., and Espevik T. The complex pattern of cytokines in serum from patients with meningococcal septic shock: association between interleukin 6, interleukin 1, and fatal outcome. J. Exp. Med. 169, 333, 1989.
5. Kitchens R. L., Wang P. Y., and Munford R. S. Bacterial lipopolysaccharide can enter monocytes via two CD14-dependent pathways. J. Immunol. 161, 5534, 1998.
6. Bonifacino J. S., and Traub L. M. Signals for sorting of transmembrane proteins to endosomes and lysosomes. Annu. Rev. Biochem. 72, 395, 2003.
7. Pickart C. M. Ubiquitin enters the new millennium. Mol. Cell 8, 499, 2001.
8. Haglund K., Sigismund S., Polo S., Szymkiewicz I., Di Fiore P. P., and Dikic I. Multiple monoubiquitination of RTKs is sufficient for their endocytosis and degradation. Nat. Cell Biol. 5, 461, 2003.
9. Huang F., Kirkpatrick D., Jiang X., Gygi S., and Sorkin A. Differential regulation of EGF receptor internalization and degradation by multiubiquitination within the kinase domain. Mol. Cell 21, 737, 2004.
10. Raiborg C., Bache K. G., Gillooly D. J., Madshus I. H., Stang E., and Stenmark H. Hrs sorts ubiquitinated proteins into clathrin-coated microdomains of early endosomes. Nat Cell Biol. 4, 394, 2002.
11. Stenmark H., Aasland R., and Driscoll P. C. The phosphatidylinositol 3-phosphate-binding FYVE finger. Febs Letters 513, 77, 2002.
12. Raiborg C., Wesche J., Malerod L., and Stenmark H. Flat clathrin coats on endosomes mediate degradative protein sorting by scaffolding Hrs in dynamic microdomains. J. Cell Sci. 119, 2414, 2004.
13. Husebye H., Halaas O., Stenmark H., Tunheim G., Sandanger O., Bogen B., Brech A., Latz E., and Espevik T. Endocytic pathways regulate Toll-like receptor 4 signaling and link innate and adaptive immunity. EMBO J. 25, 683, 2006.

14. Brissoni B., Agostini L., Kropf M., Martinon F., Swoboda V., Lippens S., Everett H., Aebi N., Janssens S., Meylan E., Felberbaum-Corti M., Hirling H., Gruenberg J., Tschopp J., and Burns K. Intracellular trafficking of interleukin-1 receptor I requires Tollip. Curr. Biol. 16, 2265, 2004.

15. Latz E., Visintin A., Lien E., Fitzgerald K. A., Monks B. G., Kurt-Jones E. A., Golenbock D. T., and Espevik T. Lipopolysaccharide rapidly traffics to and from the Golgi apparatus with the toll-like receptor 4-MD-2-CD14 complex in a process that is distinct from the initiation of signal transduction. J. Biol. Chem. 277, 47834, 2002.

16. Damke H., Baba T., Warnock D. E., and Schmid S. L. Induction of mutant dynamin specifically blocks endocytic coated vesicle formation. J. Cell Biol. 127, 915, 1994.

17. Motley A., Bright N. A., Seaman M. N. J., and Robinson M .S. Clathrin-mediated endocytosis in AP-2-depleted cells. J. Cell Biol. 162, 909, 2003.

18. Pelkmans L., Fava E., Grabner H., Hannus M., Habermann B., Krausz E., and Zerial M. Genome-wide analysis of human kinases in clathrin- and caveolae/raft-mediated endocytosis. Nature 436, 78, 2005.

19. Altschuler Y., Barbas S. M., Terlecky L. J., Tang K., Hardy S., Mostov K. E., and Schmid S. L. Redundant and distinct functions for dynamin-1 and dynamin-2 isoforms. J. Cell Biol. 143, 1871, 1998.

20. Stang E., Blystad F. D., Kazazic M., Bertelsen V., Brodahl T., Raiborg C., Stenmark H., and Madshus I. H. Cbl-dependent ubiquitination is required for progression of EGF receptors into clathrin-coated pits. Mol Biol. Cell 15, 3591, 2004.

21. Gillooly D. J., Morrow I. C., Lindsay M., Gould R., Bryant N. J., Gaullier J. M., Parton R. G., and Stenmark H. Localization of phosphatidylinositol 3-phosphate in yeast and mammalian cells. EMBO J. 19, 4577, 2000.

22. Petiot A., Faure J., Stenmark H., and Gruenberg J. PI3P signaling regulates receptor sorting but not transport in the endosomal pathway. J. Cell Biol. 162, 971, 2003.

23. Stenmark H., Aasland R., Toh B. H., and D'Arrigo A. Endosomal localization of the autoantigen EEA1 is mediated by a zinc-binding FYVE finger. J. Biol. Chem. 271, 24048, 1994.

24. Raiborg C., Bache K. G., Mehlum A., Stang E., and Stenmark H. Hrs recruits clathrin to early endosomes. EMBO J. 20, 5008, 2001.

25. Scoles D. R., Qin Y., Nguyen V., Gutmann D. H., and Pulst S. M. Hrs inhibits EGF receptor signaling in the RT4 rat schwannoma cell line. Biochem. Biophys. Res. Commun. 335, 385, 2005.

26. Elbashir S. M., Harborth J., Lendeckel W., Yalcin A., Weber K., and Tuschl T. Duplexes of 21-nucleotide RNAs mediate RNA interference in cultured mammalian cells. Nature 411, 494, 2001.

27. Harborth J., Elbashir S. M., Bechert K., Tuschl T., and Weber K. Identification of essential genes in cultured mammalian cells using small interfering RNAs. J. Cell Sci. 114, 4557, 2001.

5 The Role of Small RhoGTPases in TLR Signaling

Monica Ruse and Ulla G. Knaus

CONTENTS

5.1 INTRODUCTION

Small RhoGTPases are molecular switches that are involved in a variety of cellular processes, including reorganization of the actin cytoskeleton and microtubule network, cell cycle progression, transcriptional control, and immune responses such as phagocytosis.[1] In resting cells, the RhoGTPases Rac, Cdc42, and RhoA are sequestered in the cytosol by the guanosine 5 diphosphate (GDP) dissociation inhibitor, RhoGDI, in their inactive, GDP-bound form. Exposure of cells to pathogens or soluble stimuli including pathogen-associated molecular patterns (PAMPs) or growth factors initiates the activation of guanine nucleotide exchange factors (GEFs) that will cause the exchange of GDP to GTP and the dissociation of RhoGDI from RhoGTPases.[2] As a result, RhoGTPases are fully active and can bind to specific effector proteins to modulate cellular events. Cell type, as well as stimulus-specific activation of RhoGTPases, is ensured by the activation of only a limited subset of GEFs and by the interaction with selective effectors downstream.

The role of RhoGTPases in TLR-mediated signaling pathways just started to emerge. One challenge consists in placing a specific RhoGTPase in particular space and time coordinates along TLR-signaling pathways. Most of the studies to date relied on the use of dominant negative constructs and measure the output far downstream from the originating stimulus (e.g., cytokine production in response to bacterial components). We reported earlier an association between TLR2 and Rac1

73

in epithelial and monocytic cells.[3] In this case, heat-inactivated *Staphylococcus aureus* caused the recruitment of Rac1 together with PI-3 kinase to the cytoplasmic tail of the TLR2 receptor. The proper assembly of this complex was critical for subsequent activation of the NF-κB pathway. Others have placed Rac1 further downstream from TLR2 or TLR4, its activation being important for the initiation of proinflammatory immune responses such as cytokine or nitric oxide production.[4,5] RhoA is also rapidly activated by TLR2 and TLR4 ligands.[6,7] In monocytic cells, it can form a complex with PKCζ that is required for NF-κB-dependent gene transcription.[7] Several reports indicated the activation of Cdc42 in response to LPS or gram-positive bacterial components; however, the biological relevance of this event is still unclear.[3,8]

Here we describe methods used in our laboratory to address the role of the commonly studied RhoGTPases Rac1, Rac2, Cdc42, and RhoA in TLR-signaling pathways. The protocols for RhoGTPase immunoprecipitation or activation are very detailed, since they can be applied in various cell types regardless of the stimulus. However, in the case of methods intended for modulation of endogenous RhoGTPase expression or activity, we offer a general overview of the current methods, as they are highly dependent on the cell type under investigation.

5.2 IMMUNOPRECIPITATION AND WESTERN BLOT

Although some studies describe the immunoprecipitation of endogenous Rac1 or RhoA from cell lysates, the currently available antibodies perform poorly with often irreproducible results. In our experience, only the overexpressed forms of N-terminal epitope-tagged RhoGTPases are successfully isolated from whole cell lysates. Therefore, the following protocol refers to immunoprecipitation of Myc-tagged, wild-type Rac1 from transiently transfected HEK-293 cells. This protocol can be applied to other N-terminal tags as well as mutant forms of RhoGTPases. If a time course of stimulus-dependent RhoGTPase immunoprecipitation is desired, single time points might have to be handled separately, as rapid cycling of RhoGTPases from GDP to GTP and back to GDP necessitates fast and accurate handling of time points (30 seconds, 1 minute, etc.).

Protocol 5.1: Immunoprecipitation and Western Blot of Myc-Tagged Rac1 Protein

1. Cells transfected with Myc-Rac1wt (100–200 ng/35 mm dish) for 24 hours are washed once with cold phosphate buffer saline (PBS). On ice, add the appropriate volume of ice-cold lysis buffer (for a 35 mm dish, 75–85% confluent, add 100 μl lysis buffer). Collect by scraping the dish, and centrifuge the lysate at 13,000 rpm for 5 minutes at 4 °C in a cooling tabletop microcentrifuge.
2. Save the supernatant as the total cell lysate pool. Perform protein assay of your choice.

3. Incubate 100–300 µg of total cell lysates with 1–2 µg of anti-Myc antibody (clone 9E10) for 2 hours at 4 °C with gentle rotation of the tube.
4. Add 20–30 µl of ProteinG bead slurry (1:1 suspension in H_2O). Incubate with gentle rotation at 4 °C for 1 hour.
5. Short spin at 3,000 rpm for 30 seconds. Wash four times with cold lysis buffer.
6. Resuspend beads in 50 µl of 4× concentrated Laemmli sample buffer. Elute proteins by heating at 95 °C for 3 minutes and vortex.
7. Short spin at 14,000 rpm for 30 seconds. Save supernatant.
8. Load 20 µl of supernatant on 15% SDS-PAGE gel. Use 5–10% of the total cell lysates as a positive control for immunoblotting.
9. Transfer to nitrocellulose membranes with standard methods.
10. Block nitrocellulose membranes with 5% nonfat dried milk (by weight, or w/v) in blocking buffer for 1 hour at room temperature (RT).
11. Wash membranes several times with membrane-washing buffer at room temperature.
12. Incubate with anti-Myc antibody (preferably with polyclonal antibody) in antibody dilution buffer for 2 hours at RT.
13. Wash membranes several times with membrane-washing buffer.
14. Incubate membranes with the appropriate HRP-coupled secondary antibody in antibody dilution buffer for 45 minutes at room temperature.
15. Wash membranes several times with membrane-washing buffer.
16. Remove excess washing buffer, and incubate with ECL reagents following manufacturer's instructions.
17. Expose to X-ray film for 30 seconds up to 5 minutes. Adjust exposure time based on signal intensity.

Reagents

1. Stock solutions of protease and phosphatase inhibitors. Prepare 100 mM phenylmethanesulphonyl fluoride (PMSF) stock solution in isopropanol. Prepare 10 mM leupeptin, 1 M NaF, and 100 mM Na_3VO_4 in distilled water.
2. Aprotinin (Sigma-Aldrich, St. Louis, Missouri).
3. 10× Tris-buffered saline (10×TBS).
4. Lysis buffer: 25 mM Tris-HCl (pH 7.5), 1 mM EDTA, 5 mM $MgCl_2$, 0.1 mM DTT, 0.1 mM EGTA, 100 mM NaCl, 10% glycerol, 1% Nonidet P-40, 1 mM PMSF, 200 µM leupeptin, 1µg/ml aprotinin, 10 mM NaF, and 2 mM Na_3VO_4. Add protease and phosphate inhibitors into the ice-cold buffer shortly before the procedure.
5. Protein G Sepharose (GE Healthcare Bio-Sciences, Piscataway, NJ).
6. Laemmli sample buffer (1×): 2% sodium dodecylsulfate (SDS), 10% glycerol, 2% β-mercaptoethanol, 60 mM Tris-HCl (pH 6.8), and trace of bromophenol blue.
7. Suggested antibodies: Myc monoclonal antibody (9E10; Upstate Biotechnology, Lake Placid, NY), Myc polyclonal antibody (Santa Cruz Biotechnology,

Santa Cruz, CA), Rac1 monoclonal antibody (Cytoskeleton, Denver, CO; or Upstate Biotechnology), Cdc42 polyclonal antibody (Santa Cruz Biotechnology), and RhoA monoclonal antibody (Santa Cruz Biotechnology).

8. Nitrocellulose membrane (Bio-Rad Laboratories, Hercules, CA).
9. Blocking buffer: 3% (w/v) BSA, 10% (by volume, or v/v) goat serum, and 0.02% sodium azide in TBS.
10. Membrane-washing buffer: 0.5% (v/v) Tween-20 in TBS (TBS-T).
11. Antibody dilution buffer: 1% (w/v) BSA, 3% (v/v) goat serum, and 0.5% (v/v) Tween-20 in TBS.
12. Secondary HRP-conjugated antibodies (e.g. Southern Biotech, Promega).
13. Enhanced chemiluminescence (ECL) reagents (Pierce Biotechnology, Rockford, IL).

5.3   AFFINITY-BASED RHOGTPASE ACTIVATION ASSAYS

This method relies on the premise that active, GTP-bound forms of RhoGTPases are recognized by selective high-affinity domains that are integral parts of downstream effectors. The Rac/Cdc42–specific binding probe is based on a domain within the N-terminal regulatory region of p21-kinase 1 (Pak1). This region, encompassing the amino acids 67 to 150, is also referred to as p21-binding domain, or PBD. The PBD will bind active Rac1, Rac2, as well as Cdc42.[9] For selective isolation of active Cdc42, a probe derived from Wiskott Aldrich Syndrome Protein (WASP) is used.[10] In addition, a domain derived from the Rho effector Rhotekin (amino acids 7 to 89) can be employed for selective binding of active RhoA.[11] All of the probes are fused to glutathione-S transferase and coupled to Glutathione Sepharose 4B beads (GE Healthcare Bio-Sciences, Piscataway, NJ), which will permit the precipitation of a complex formed by active RhoGTPase and the corresponding GST-effector-binding domain. The precipitated RhoGTPase will be visualized on a Western blot using specific Rac, Cdc42, or Rho antibodies. Our protocol describes the preparation of the GST-PBD probe as well as the detection of active Rac1 utilizing this probe. A representative experiment depicting Cdc42 activation by polyIC in bone marrow–derived macrophages is shown (Figure 5.1). The preparation and use of WASP or Rhotekin-based GTPase activation probes follow the same principle and can be found elsewhere.[10,11] However, for those interested in shortening the preparatory steps, small GTPase activation assay kits are now commercially available from Cytoskeleton (Denver, CO) or Upstate Biotechnology (Lake Placid, NY).

Protocol 5.2: Expression of GST-PBD Fusion Protein in Bacteria

1. Competent *E. coli* strain BL21(DE3)LysE is transformed with pGEX-PBD (aa 67–150 of hPak1) plasmid by standard procedures and grown on Luria-Bertani broth (LB) agar plates with 100μg/ml ampicillin. The LB/ampicillin

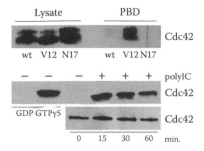

FIGURE 5.1 TLR3-Mediated Cdc42 Activation in Murine Macrophages

Note: Specificity of PBD beads for active Cdc42 was tested using cell lysates derived from HeLa cells expressing wild-type (see source line below for reference), constitutively active (V12), and dominant negative (N17) Cdc42 followed by immunoblotting (first panel). Bone marrow–derived macrophages were left unstimulated or were stimulated with 10 μg/mL polyIC for the indicated time points (second panel). Cells were lysed and Cdc42-GTP was bound to the GST-PBD probe followed by immunoblot analysis. As additional control, one-tenth of the unstimulated cell lysate was loaded with GTPγS or GDP. The last panel shows a Cdc42 immunoblot using one-tenth of the total cell lysate for each time point.

Source: King, C. C., Gardiner, E. M., Zenke, F. T., Bohl, B. P., Newton, A. C., Hemmings, B. A., and Bokoch, G. M. p21-activated kinase (PAK1) is phosphorylated and activated by 3-phosphoinositide-dependent kinase-1 (PDK1). *J. Biol. Chem.* 275, 41201, 2000.

culture is inoculated with an individual bacterial colony and grown until the log phase is achieved. An aliquot of this culture is used for bacterial stock preparation. Dilute the stock culture 1:1 with sterile 30% (v/v) glycerol, snap-freeze in liquid N_2, and store at −80 °C.

2. From the frozen stock or a streaked plate, a 50 ml LB/ampicillin culture is inoculated and grown overnight with shaking at 37 °C. The next morning, 25 ml of this culture are transferred to 1 L of LB/ampicillin broth, and bacterial growth is continued until the OD_{600nm} reaches 0.6–0.8 (usually 2 hours). Save 50 μl of bacterial culture for final SDS-PAGE analysis (this will be the uninduced sample).

3. Induce transcription with 0.8 mM IPTG, and grow for 3 hours at 30 °C. Save 50 μl of this culture for final SDS-PAGE analysis (this will be the induced sample).

Protocol 5.3: Preparation of GST-PBD Probe

1. Prepare the *E. coli* lysate by spinning down the culture for 10 minutes at 2,300g at 4 °C. Wash the pellet once with cold PBS followed by centrifugation, and resuspend the pellet in 10 ml cold bacterial lysis buffer. Incubate on ice for 10 minutes, and sonicate afterwards three times for 30 seconds on ice without foaming (output 70%). If necessary, the *E. coli* pellet can be stored after the PBS wash at −80 °C. In this case, addition of 1 mg/ml lysozyme

and 20 μg/ml DNase I is recommended for efficient bacterial lysis and DNA shredding.

2. Clear the lysate by centrifugation at 9,300g for 15 minutes at 4°C. Collect the supernatant and save 50 μl for final SDS-PAGE analysis (this will be the GST-PBD total sample).

3. Prepare Glutathione Sepharose 4B beads (GE Healthcare Bio-Sciences, Piscataway, NJ). Take the equivalent of 1 ml dry beads (for 1L *E. coli* culture) and centrifuge 400g for 5 minutes at 4 °C in a tabletop microcentrifuge. Wash the beads two times with 10 ml H₂O and two times with 5 ml bacterial lysis buffer.

4. Combine washed Glutathione Sepharose 4B beads with *E. coli* supernatant from step 2 and incubate for 1–2 hours (recommended) or overnight at 4 °C while inverting.

5. Centrifuge 400g for 5 minutes at 4 °C, and save 50 μl of supernatant for final SDS-PAGE analysis (this will be the GST-PBD unbound protein sample).

6. Wash beads five times with ice-cold bacterial lysis buffer.

7. Aliquot beads in washing buffer containing 10% (v/v) glycerol (1:1 slurry), and store at −80 °C. Determine the protein concentration of GST-PBD bound to beads with the commercially available protein assays.

8. Analysis of the GST-PBD purification process by SDS-PAGE. Analyze 5 μl of samples from uninduced, induced, GST-PBD total, GST-PBD unbound, and final purified GST-PBD bound to beads aliquots (use 5 and 10 μl of 1:1 bead slurry) by SDS-PAGE and Coomassie blue staining to determine protein induction, amounts of soluble and bound protein, and purity of the final GST-PBD probe.

9. Analysis of GST-PBD beads for efficient RhoGTPase-binding capacity. This assay relies on the ability of mutant RhoGTPases to be locked in vitro into the GDP or GTP state. Wild-type, dominant negative, and constitutively active RhoGTPases containing N-terminal epitope tags are transiently transfected into model cell lines (HEK293, HeLa, etc.). After 24 hours, total cell lysates are prepared, and aliquots are incubated with GST-PBD beads as described in detail in Protocol 5.4. Constitutively active RhoGTPases Rac1/2 and Cdc42 will bind to the probe, while the dominant negative mutants of these GTPases will not.

Protocol 5.4: Detection of GTP-Bound Rac Utilizing GST-PBD Probe

1. The cells are stimulated to generate activated Rac (GTP-bound Rac). For suspension cells, an aliquot of cells is removed at certain time points, and the stimulation is stopped by adding this aliquot to 2–4× concentrated ice-cold PBD binding buffer. For adherent cells, the stimulation medium is removed at certain time points, 1× ice-cold PBD binding buffer is added, and the cells are scraped from the plates on ice. The resulting total cell lysates are kept on

ice for several minutes, vortexed, and clarified by centrifugation at 9,300g for 10 minutes at 4 °C. The supernatants should be placed on ice for immediate use.

2. For each affinity precipitation sample, an equal number of cells or the same amount of total cell protein should be mixed with the GST-PBD probe. The maximal final volume for each affinity precipitation sample is 500 µl, and if required, the samples are diluted in PBD binding buffer. Save one-tenth to one-twentieth of the original sample for immunoblot determination of total Rac content in the lysate.

3. Add 10 µg of purified GST-PBD beads to the samples, and incubate for 1 hour at 4 °C while inverting. Centrifuge the samples at 400g for 2 minutes at 4 °C, aspirate the supernatant, and wash 3 times with PBD binding buffer.

4. Add 4× Laemmli sample buffer, and heat at 95 °C for 3 minutes. Load samples onto a 12–15% SDS-PAGE gel. Include control reactions (see below). Transfer to nitrocellulose or PVDF membrane, and stain the membrane with Ponceau S reagent to ensure equal distribution of the GST-PBD beads in different samples. After blocking of the membrane, perform immunoblot analysis with the appropriate Rac antibodies.

5. Positive and negative controls for the assay should be performed with lysates from unstimulated cells, in which endogenous Rac is loaded in vitro with either GDP (negative control) or GTPγS (positive control). As a positive control (GTP-bound form), simultaneously add GTPγS at a final concentration of 100 µM and EDTA at a final concentration of 10 mM to the cell lysates prepared in cell lysis buffer. Incubate for 15 minutes at 30 °C and add on ice $MgCl_2$ to a final concentration of 60 mM to stop the nucleotide exchange. As a negative control (GDP-bound form), simultaneously add GDP at a final concentration of 1 mM and EDTA at a final concentration of 10 mM to cell lysates prepared in cell lysis buffer. Incubate for 15 minutes at 30 °C and add on ice $MgCl_2$ to a final concentration of 60 mM to stop the nucleotide exchange. The GTPγS- or GDP-loaded samples should be kept on ice and used within the next 15–30 minutes by following the procedure for the binding assays described in step 3.

6. On a special note, the affinity-based RhoGTPase activation assay will not provide the ratio of activated versus nonactivated Rac1. For this, a GTP- or GDP-loading method is performed. Briefly, the cells are incubated with orthophosphate prior to stimulation and the RhoGTPase is precipitated with specific antibodies if these are available. The immunocomplexes are loaded on a thin-layer chromatography (TLC) plate, and the GDP or GTP spots are further separated by TLC. Quantification is performed using a Phosphorimager (Molecular Dynamics, Piscataway, NJ). The success of this method relies on efficient immunoprecipitation of RhoGTPases from total cell lysates.

Reagents

1. The Rac/Cdc42–binding domain of p21-activated kinase (Pak1; GST-PBD, aa 67–150) cloned into pGEX bacterial expression vector (GE Healthcare Bio-Sciences, Piscataway, NJ).
2. Luria broth.
3. Ampicillin (final concentration 100 µg/ml).
4. Isopropyl-b-D-thio-galactopyranoside (IPTG).
5. Bacterial lysis buffer: 50 mM Tris-HCl (pH 7.5), 150 mM NaCl, 5 mM MgCl$_2$, 1 mM EDTA, 1 mM dithiothreitol (DTT), 1 mM phenylmethylsulfonyl fluoride (PMSF), and 1 µg/ml aprotinin (Sigma-Aldrich, St. Louis, Missouri). Add DTT, PMSF, and aprotinin to the freshly made, ice-cold buffer prior to use.
6. Bacterial wash buffer: 50 mM Tris-HCl (pH 8.0), 150 mM NaCl, 5 mM MgCl$_2$, 1 mM DTT, 1 mM PMSF, and 1 µg/ml aprotinin. Add DTT, PMSF, and aprotinin to the freshly made, ice-cold buffer prior to use.
7. Glutathione Sepharose 4B beads (GE Healthcare Bio-Sciences, Piscataway, NJ).
8. Cell lysis buffer: 50 mM Tris-HCl (pH 7.5), 150 mM NaCl, 5 mM MgCl$_2$, 1 mM EDTA, 1% NP-40, 10% glycerol, 1 mM DTT, 1 mM PMSF, 5 µg/ml aprotinin, 1 µg/ml leupeptin, and 2 mM sodium orthovanadate. Add DTT, protease, and phosphatase inhibitors to the freshly made, ice-cold buffer prior to use.
9. PBD binding buffer: 25 mM Tris-HCl (pH 7.5), 40 mM NaCl, 30 mM MgCl$_2$, 1% NP-40, 1 mM DTT, 1 mM PMSF, 5 µg/ml aprotinin, and 1 µg/ml leupeptin. Add DTT and protease inhibitors to the freshly made, ice-cold buffer prior to use.
10. Phosphate buffer saline.
11. GTPγS tetralithium salt (10220647001; Roche Applied Science, Basel, Switzerland) and GDP Tris salt (G7252; Sigma-Aldrich, St. Louis, Missouri).
12. Laemmli sample buffer (1×): 2% sodium dodecylsulfate (SDS), 10% glycerol, 2% β-mercaptoethanol, 60 mM Tris-HCl (pH 6.8), and trace of bromophenol blue.
13. Ponceau S solution: 0.2% Ponceau S in 3% trichloroacetic acid.
14. Antibodies: anti-Rac1 mAb (Cytoskeleton, Denver, CO; or clone 23A8 from Upstate Biotechnology, Lake Placid, NY) and anti-Rac2 pAb (Upstate Biotechnology).

5.4 CHANGES IN RHOGTPASE PROTEIN EXPRESSION OR ACTIVITY

Recently, progress has been made in gene-targeting individual RhoGTPases to discern their physiological in vivo roles and involvement in signaling pathways. Embryonic lethality associated with deficiency in Rac1, Cdc42, and RhoA requires a conditional knockout approach. Due to these constraints, immunological phenotypes have been described only for Rac2 and Rac1.[12-15] Both GTPases play distinct roles in neutrophil chemotaxis and pathogen phagocytosis.[12-17] Downregulation of RhoGT-Pases in fibroblasts or epithelial cells via small interfering RNAs (siRNAs) has been accomplished and offers a potent alternative for cell culture studies in these cell types. In both cases, adaptive compensation by an increase in related RhoGTPases is an inherent concern. Additionally, RhoGTPases can act sequentially in certain signaling pathways. Thus, the absence of the targeted GTPase will affect the activity of another GTPase. Due to these special features, a correct understanding of RhoGTPase function by deletion or knockdown of RhoGTPases requires analysis of expression and activation of related RhoGTPases in the context of the study.

Constitutive active or dominant negative constructs for RhoGTPases were traditionally used as tools to delineate the contribution of a specific GTPase in various cellular processes. Immortalized cell lines are amenable to lipid-based transfection methods (i.e., FuGENE and Lipofectamine), whereas the manipulation of RhoGT-Pase activity in primary immune cells such as macrophages or neutrophils relies on protein transduction or infection by lentivirus or retrovirus. Providing comprehensive protocols for each of these methods is beyond the scope of this review, since each method needs to be optimized depending on the cell type utilized. However, we will briefly outline selected methods and indicate the relevant references for further analysis. We have successfully utilized transient transfection of HEK-293 cells with Lipofectamine Plus reagent (Invitrogen, Carlsbad, CA) or of THP-1 cells with FuGENE (Roche Applied Science, Basel, Switzerland) to study the requirement of specific RhoGTPases in TLR-mediated signaling pathways.[3,7]

5.4.1 BioPORTER®

Advances in protein transduction methods allowed the study of RhoGTPase function in primary immune cells. BioPORTER® (Gene Therapy Systems, San Diego, CA) is a lipid-based approach that ensures the delivery of active, purified recombinant proteins into the living cells. The advantages of this method are manifold: optimum delivery in 3 to 4 hours after incubation, efficient in many cell types (including primary cells), and noncytotoxic. The BioPORTER® reagent was used to transduce the Cdc42-binding domain of the Cdc42 effector Wiskott Aldrich syndrome protein (WASP-CRIB) or to deliver the constitutive active form of Cdc42 (Q61L) into human neutrophils.[8,18] In another study, the role of Rac1 and RhoA in human neutrophil chemotaxis was delineated by delivering the corresponding GTPase mutants as a mixture with the BioPORTER® reagent.[19]

Protocol 5.5: Transduction of Human Neutrophils with RhoGTPases Using the BioPORTER® Reagent

1. Dissolve each tube of the dried BioPORTER® reagent in 250 µl chloroform. Vortex for 10–20 seconds at maximal speed.

2. Immediately aliquot 2 μl of the dissolved BioPORTER® reagent into Eppen-dorf tubes. Work fast, since the chloroform evaporates quickly and this can create variability in the final concentration of the reagent. Place the tubes under a laminar flow hood with the caps open, and dry overnight at room temperature.

3. Dried BioPORTER® can be stored at –20 °C for at least 1 year.

4. 100 μl of PBS containing 10–30 μg of recombinant RhoGTPase can be used to rehydrate the BioPORTER® reagent. Incubate for 5 minutes at room temperature and vortex.

5. Add 3×10^6 neutrophils in 900 μl KRGH buffer (see under "Reagents," below), and invert the mixture for 1–2 hours at room temperature.

6. Pellet the neutrophils at 500g for 3 minutes. Resuspend in the culture medium of your choice.

7. Neutrophils are ready to be used in subsequent experiments for stimulation with TLR ligands or phagocytosis studies.

Reagents

1. BioPORTER® reagent (Gene Therapy Systems, San Diego, CA).

2. Phosphate-buffered saline and chloroform.

3. Purified recombinant RhoGTPase proteins.

4. KRGH buffer: 25 mM Hepes (pH 7.4), 1.2 mM KH_2PO_4, 118 mM NaCl, 4.7 mM KCl, 1 mM $MgSO_4$, 1 mM $CaCl_2$, and 5.5 mM dextrose.

5.4.2 TAT-MEDIATED RHOGTPASE PROTEIN TRANSDUCTION

Another method that allows efficient protein delivery in a variety of cells is the HIV Tat-mediated protein transduction.[20] The histidine-tagged protein of interest is fused to a polybasic domain of Tat protein (amino acids 11–36) and is purified by affinity chromatography after expression in *E. coli*. Crossing the cellular membranes by the Tat-fusion protein is a rapid, receptor-independent process (5–10 minutes), and the method can be used for virtually any cell type. We utilized this tool in studying the role of p21-activated kinase (Pak) in human neutrophils. By delivering a Pak inhibitory domain– (PID-) Tat-fusion protein to human neutrophils, we demonstrated the requirement of Pak for efficient NADPH oxidase-dependent superoxide generation.[21] Zhou et al. showed that Pak1 is important for Akt-stimulated cell migration by transducing HeLa and Rat1 cells with a peptide derived from the Pak1 adapter Nck fused to Tat.[22]

Protocol 5.6: Tat-Mediated RhoGTPase Transduction

1. Clone the RhoGTPase of interest in a bacterial expression vector containing the Tat sequence. We used successfully the pTat-HA expression vector from

Dr. S. Dowdy (University of California, San Diego). The green fluorescent protein (GFP) sequence can be added to the protein of interest when cloning to monitor the transduction efficiency. Alternatively, purified Tat-RhoGT-Pases can be labeled with fluorescent labels such as fluorescein isothiocyanate (FITC) to monitor uptake. Labeling should be avoided in functional studies.

2. Tat-RhoGTPase plasmids are used for transformation of *E. coli* BL21(DE3) LysS competent cells. Once the positive clones are identified, a large-scale bacterial culture is required. 100 ml of the bacterial culture expressing Tat-RhoGTPase is grown overnight. Next morning, use the entire culture to inoculate 1L of LB, and shake at 37 °C. Add 0.5 mM of IPTG after 2–3 hours, and shake up the mixture for 5–10 hours at 30 °C.

3. Centrifuge the culture, and wash the pellet with PBS. Resuspend the pellet in 10 ml bacterial lysis buffer, and sonicate on ice until solution is turbid.

4. Clarify solution by centrifugation at 5,000 rpm for 10 minutes at 4 °C. Save the supernatant.

5. Preequilibrate the Ni-NTA column in bacterial lysis buffer containing 10 mM imidazole (equilibration buffer). Add 10 mM imidazole to the supernatant, and then apply it to the column. Before adding the supernatant to the column, save an aliquot for SDS-PAGE analysis. Save also an aliquot of the flow-through.

6. Wash the column with equilibration buffer.

7. Elute Tat-fusion protein by the gradual addition of 5 ml of bacterial lysis buffer containing increasing amounts of imidazole: 100 mM, 250 mM, 500 mM, and 1 M. Save each fraction for subsequent SDS-PAGE analysis. Pool together the fractions containing the Tat-fusion protein. These fractions can be frozen for further purification.

8. The purified Tat-fusion proteins are in 8 M urea-containing buffer. For further purification and buffer exchange, use either ion exchange or a gel filtration column.

9. Ion exchange: Set up a 5 ml ion exchange column with Mono Q resin. Equilibrate columns with ion exchange equilibration buffer. Dilute sample 1:1 in 20 mM HEPES (pH 8) and inject into the preequilibrated Mono Q column using a syringe (save an aliquot of the starting material as well as flow-through for SDS-PAGE analysis).

10. Wash the column with ion exchange equilibration buffer containing 10 mM imidazole (save an aliquot of the wash).

11. Elute the Tat-fusion protein in a single step with 1–2 M NaCl (elution buffer).

12. Analyze all the fractions on an SDS-PAGE gel, and pool together the fractions containing Tat-fusion protein.

13. Following step 11, the Tat-fusion proteins are eluted in a buffer with a high salt content. The sample can be desalted by gel filtration.

14. Gel filtration: Sephadex G-25 column is equilibrated in sterile PBS. Save aliquots of the starting material, flow-through, and collected fractions to analyze on an SDS-PAGE gel.

Reagents

1. BL21(DE3) LysS competent cells (Invitrogen, Carlsbad, CA)
2. Luria broth
3. Isopropyl-b-D-thio-galactopyranoside (IPTG)
4. Bacterial lysis buffer: 20 mM HEPES (pH 8.0), 100 mM NaCl, and 8 M urea (add up to 500 mM NaCl to improve protein recovery if necessary)
5. Imidazole
6. Ion exchange equilibration buffer: 20 mM HEPES (pH 8.0) and 50 mM NaCl
7. Ni-NTA resin (Qiagen, Valencia, CA)
8. Mono Q resin (GE Healthcare Bio-Sciences, Piscataway, NJ)
9. Elution buffer: 20 mM HEPES (pH 8.0) and 1–2 M NaCl
10. Phosphate-buffered saline
11. Sephadex G-25 column (GE Healthcare Bio-Sciences, Piscataway, NJ)

5.4.3 VIRAL INFECTION

Infection of primary immune cells with lentivirus or retrovirus that carries GTPase constructs can be employed as a valuable approach to overcome the failure to express DNA constructs by traditional lipid-based methods or to introduce short hairpin RNA (shRNA) constructs. However, the potential pitfalls involve the undesired activation of these cells by Toll-like receptors responsive to viral components (e.g., TLR3). Control experiments and optimization are required to ensure the specificity of the observed effects.

As a general remark, many of the methods detailed above rely on the delivery of constitutively active or inhibitory dominant negative RhoGTPases. However, great care has to be taken in the interpretation of these results since dominant negative constructs can affect TLR-activated GEFs that are shared by several GTPases, or constitutive active proteins might not be subjected to the same spatial-temporal regulation such as wild-type proteins. These methods are best complemented by the use of RNAi or shRNA, GTPase inhibitors (i.e., NSC23766, and C3 toxin), and, when possible, studies in RhoGTPase-deficient mice.

ACKNOWLEDGMENTS

We would like to thank S. Davanture for technical help and K. Schreiber for administrative assistance. This work was supported by NIH NIAID grants to Ulla G. Knaus.

REFERENCES

1. Jaffe, A. B., and Hall, A. RHO GTPASES: biochemistry and biology. Annu. Rev. Cell Dev. Biol. 21, 247, 2005.
2. Bokoch, G. M. Regulation of innate immunity by Rho GTPases. Trends Cell Biol. 15, 163, 2005.

3. Arbibe, L., Mira, J. P., Teusch, N., Kline, L., Guha, M., Mackman, N., Godowski, P. J., Ulevitch, R. J., and Knaus, U. G. Toll-like receptor 2-mediated NF-kappa B activation requires a Rac1-dependent pathway. Nat. Immunol. 1, 533, 2000.
4. Schmeck, B., Huber, S., Moog, K., Zahlten, J., Hocke, A. C., Opitz, B., Hammerschmidt, S., Mitchell, T. J., Kracht, M., Rosseau, S., Suttorp, N., and Hippenstiel, S. Pneumococci induced TLR- and Rac1-dependent NF-kappaB-recruitment to the IL-8 promoter in lung epithelial cells. Am. J. Physiol. Lung Cell Mol. Physiol. 290, L730, 2006.
5. Liu-Bryan, R., Pritzker, K., Firestein, G. S., and Terkeltaub, R. TLR2 signaling in chondrocytes drives calcium pyrophosphate dihydrate and monosodium urate crystal-induced nitric oxide generation. J. Immunol. 174, 5016, 2005.
6. Chen, L. Y., Zuraw, B. L., Liu, F. T., Huang, S., and Pan, Z. K. IL-1 receptor-associated kinase and low molecular weight GTPase RhoA signal molecules are required for bacterial lipopolysaccharide-induced cytokine gene transcription. J. Immunol. 169, 3934, 2002.
7. Teusch, N., Lombardo, E., Eddleston, J., and Knaus, U. G. The low molecular weight GTPase RhoA and atypical protein kinase Czeta are required for TLR2-mediated gene transcription. J. Immunol. 173, 507, 2004.
8. Fessler, M. B., Arndt, P. G., Frasch, S. C., Lieber, J. G., Johnson, C. A., Murphy, R. C., Nick, J. A., Bratton, D. L., Malcolm, K. C., and Worthen, G. S. Lipid rafts regulate lipopolysaccharide-induced activation of Cdc42 and inflammatory functions of the human neutrophil. J. Biol. Chem. 279, 39989, 2004.
9. Burbelo, P. D., Drechsel, D., and Hall, A. A conserved binding motif defines numerous candidate target proteins for both Cdc42 and Rac GTPases. J. Biol. Chem. 270, 29071, 1995.
10. Rudolph, M. G., Bayer, P., Abo, A., Kuhlmann, J., Vetter, I. R., and Wittinghofer, A. The Cdc42/Rac interactive binding region motif of the Wiskott Aldrich syndrome protein (WASP) is necessary but not sufficient for tight binding to Cdc42 and structure formation. J. Biol. Chem. 273, 18067, 1998
11. Reid, T., Furuyashiki, T., Ishizaki, T., Watanabe, G., Watanabe, N., Fujisawa, K., Morii, N., Madaule, P., and Narumiya, S. Rhotekin, a new putative target for Rho bearing homology to a serine/threonine kinase, PKN, and rhophilin in the rho-binding domain. J. Biol. Chem. 271, 13556, 1996.
12. Glogauer, M., Marchal, C. C., Zhu, F., Worku, A., Clausen, B. E., Foerster, I., Marks, P., Downey, G. P., Dinauer, M., and Kwiatkowski, D. J. Rac1 deletion in mouse neutrophils has selective effects on neutrophil functions. J. Immunol. 170, 5652, 2003.
13. Yamauchi, A., Kim, C., Li, S., Marchal, C. C., Towe, J., Atkinson, S. J., and Dinauer, M. C. Rac2-deficient murine macrophages have selective defects in superoxide production and phagocytosis of opsonized particles. J. Immunol. 173, 5971, 2004.
14. Kim, C., and Dinauer, M. C. Rac2 is an essential regulator of neutrophil nicotinamide adenine dinucleotide phosphate oxidase activation in response to specific signaling pathways. J. Immunol. 166, 1223, 2001.
15. Sun, C. X., Downey, G. P., Zhu, F., Koh, A. L., Thang, H., and Glogauer, M. Rac1 is the small GTPase responsible for regulating the neutrophil chemotaxis compass. Blood 104, 3758, 2004.
16. Abdel-Latif, D., Steward, M., Macdonald, D. L., Francis, G. A., Dinauer, M. C., and Lacy, P. Rac2 is critical for neutrophil primary granule exocytosis. Blood 104, 832, 2004.
17. Roberts, A. W., Kim, C., Zhen, L., Lowe, J. B., Kapur, R., Petryniak, B., Spaetti, A., Pollock, J. D., Borneo, J. B., Bradford, G. B., Atkinson, S. J., Dinauer, M. C., and Williams, D. A. Deficiency of the hematopoietic cell-specific Rho family GTPase Rac2 is characterized by abnormalities in neutrophil function and host defense. Immunity 10, 183, 1999.

18. Cheng, G., Diebold, B. A., Hughes, Y., and Lambeth, J. D. Nox1-dependent reactive oxygen generation is regulated by Rac1. J. Biol. Chem. 281, 17718, 2006.

19. Pestonjamasp, K. N., Forster, C., Sun, C., Gardiner, E. M., Bohl, B., Weiner, O., Bokoch, G. M., and Glogauer, M. Rac1 links leading edge and uropod events through Rho and myosin activation during chemotaxis. Blood 108, 2814, 2006.

20. Fawell, S., Seery, J., Daikh, Y., Moore, C., Chen, L. L., Pepinsky, B., and Barsoum, J. Tat-mediated delivery of heterologous proteins into cells. Proc. Natl. Acad. Sci. USA 91, 664, 1994.

21. Martyn, K. D., Kim, M. J., Quinn, M. T., Dinauer, M. C., and Knaus, U. G. p21-activated kinase (Pak) regulates NADPH oxidase activation in human neutrophils. Blood 106, 3962, 2005.

22. King, C. C., Gardiner, E. M., Zenke, F. T., Bohl, B. P. Newton, A. C., Hemmings, B. A., and Bokoch, G. M., p21-activated kinse (PAK1) is phosphorylated and activated by 3-phosphoinositide-dependent kinase-1 (PDK1). *J. Biol. Chem.* 275, 41201, 2000.

(a)

Vector Control | TLR4/Bla(a)+ TLR4/Bla(b) | MyD88/Bla(a)+ MyD88/Bla(b)

MyD88/bla(b)+ DD[MyD88]/Bla(a) | MyD88/Bla(a)+ TLR4/Bla(b) | TLR4/Bla(b)+ DD[MyD88]/Bla(a) | MyD88/Full Bla

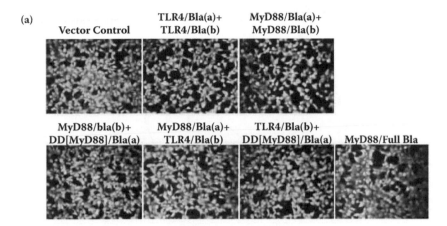

(b)

Vector Control | TLR4/Bla(a)+ TLR4/Bla(b) | MyD88/Bla(a)+ MyD88/Bla(b)

MyD88/Bla(b)+ DD[MyD88]/Bla(a) | MyD88/Bla(a)+ TLR4/Bla(b) | TLR4/Bla(b)+ DD[MyD88]/Bla(a) | MyD88/Full Bla

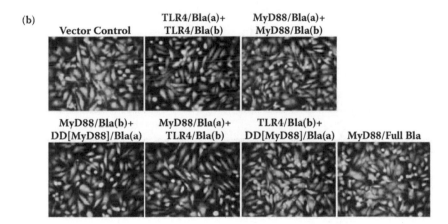

COLOR FIGURE 6.4 In Vivo β-Lactamase Complementation Analyzed by Fluorescence Microscopy

Note: HEK293 cells (A) or HeLa cells (B) were transiently co-transfected with different combinations of the β-lactamase fusion constructs (as noted) and loaded with CCF2/AM.

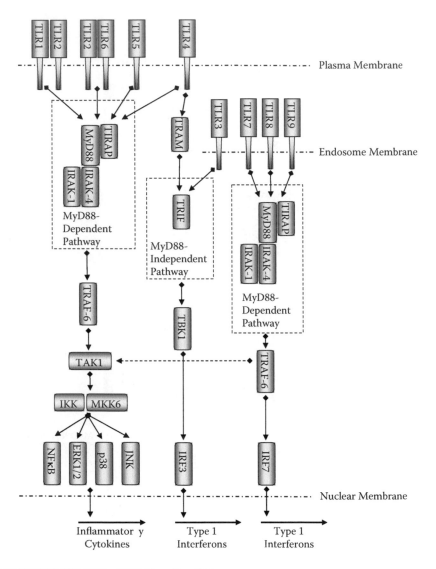

COLOR FIGURE 10.1 TLR Signaling Pathways

6 Use of Engineered β-Lactamase Fragment Complementation to Detect the Associations of Toll-Like Receptors and Signaling Adaptors

Hyun-Ku Lee and Peter S. Tobias

CONTENTS

6.1 INTRODUCTION

Toll-like receptors (TLRs) are type I transmembrane proteins and are known as pattern recognition receptors (PRRs) that can recognize conserved pathogen-associated molecular patterns (PAMPs).[1–3] As demonstrated by a number of studies, TLRs play an essential role in sensing innate immune signals that result in the production of proinflammatory and anti-inflammatory cytokines including IL-1, IL-6, IL-8, IL-10,

IL-12, TNF-α, and IFNs.[4–6] The TLR-mediated innate immune signals also stimulate the host to develop adaptive immunity through induction of major histocompatibility complex molecules as well as co-stimulatory molecules such as CD40, CD80, and CD86.[7–9] TLR activation is generally initiated through ligand-associated TLR homodimerization as demonstrated by crystal structure studies of TLR.[10,11] Cooperation among TLRs in signal transduction further describes that some TLRs also work as heterodimers.[12–16] The TLR dimerizations are sequentially followed by binding their signaling adaptors that further propagate the downstream signaling cascades. The in vivo protein-protein interactions among TLRs as well as between TLRs and their signaling adaptors can be detected by protein complementation assays (PCAs), as previously shown.[17] This chapter describes the molecular basis of this technique.

6.1.1 MAMMALIAN TLRs

TLRs consist of an N-terminal extracellular domain (ED) and a C-terminal cytoplasmic domain (CD) that is separated from ED by a short transmembrane domain (TD). The ED of each TLR contains leucine-rich repeat (LRR) motifs varying in length among TLRs, which are a common characteristic in PRRs.[18,19] The CD of TLR is homologous to that of the interleukin 1 receptor (IL-1R) because both contain the Toll/IL-1R homology (TIR) domain. Since the first mammalian Toll homologue was discovered,[20] at least 11 mammalian TLRs have been identified, and many of their ligands have been discovered[21,22] (Table 6.1). Some TLRs (including TLR-3, -7, -8, and -9) are intracellular, which are mainly located to the endosomal or lysosomal compartments.[23–25] Nevertheless, all TLRs similarly recruit their signaling adaptors when they are activated by their ligands (Table 6.1).

TABLE 6.1
TLR-Related Adaptors and Exogenous and Endogenous TLR Ligands

TLR	Adaptor	Exogenous Ligand	Endogenous Ligand
TLR2	MyD88 and TIRAP	Lipoprotein, lipoarabinomannan, and glycosylphosphati-dylinositol anchor	High-mobility group box 1 (HMGB1) proteins
TLR3	TRIF	Double-stranded RNA	mRNA
TLR4	MyD88, TIRAP, TRAM, and TRIF	LPS	Lung surfactant protein-A, hyaluronan fragment, heparan sulfate, fibrogen, and HMGB1
TLR5	MyD88	Flagellin	
TLR7	MyD88	Single-stranded RNA	
TLR8	MyD88	Single-stranded RNA	
TLR9	MyD88	Bacterial CpG DNA and hemozoin	DNA-immunoglobulin complex (DNA-IC)
TLR11	MyD88	Profilin-like molecule	

6.1.2 TLR SIGNALING ADAPTORS

TLR adaptors similarly contain a TIR domain, but they differentially interact with different TLRs, subsequently propagating signaling downstream by activating different kinases (Figure 6.1). Myeloid differentiation factor 88 (MyD88) is the most common TLR adaptor,[26] and the downstream signaling of TLRs is often divided into MyD88-dependent and MyD88-independent pathways.[27] Other adaptors include TIR domain–containing adaptor protein (TIRAP, also known as Mal), TIR domain–containing adaptor inducing IFN-β (TRIF), and TRIF-related adaptor molecule (TRAM), but all TLRs utilize MyD88 with the exception of TLR3, which uses only TRIF.[28] MyD88 contains an N-terminal death domain (DD), which is separated from a C-terminal TIR domain by a short intermediate domain (ID). The interaction of MyD88 with given receptors commonly occurs via homophilic TIR-TIR domain association, while the DD of MyD88 recruits signaling kinases resulting in a series of signaling cascades downstream. The MyD88-dependent pathway often recruits another TLR adaptor, TIRAP, especially in TLR2 or TLR4;[29,30] together, the TLRs sharing this MyD88-dependent pathway lead to NF-κB and/or AP-1-dependent proinflammatory cytokine expression (Figure 6.1). The MyD88-independent pathway is shared by only TLR3 and TLR4 and is often called the TRIF-dependent pathway because of their TRIF usage, leading to IRF3-dependent IFN-β expression; among these two TLRs, only TLR4 recruits another adaptor, TRAM, which bridges between TLR4 and TRIF[28,31] (Figure 6.1).

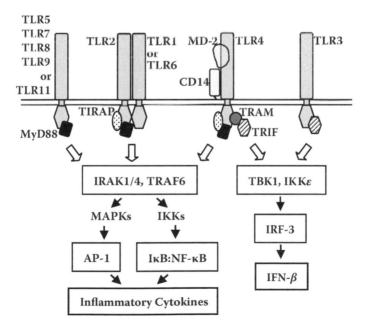

FIGURE 6.1 TLR Signal Transduction Involving Various TLR Adaptors
Note: See text for details.

6.1.3 ENZYME-BASED PCAS

Protein fragment complementation strategies to detect protein-protein interactions have long been considered by researchers. Among the proteins, enzymes are ideal sources for PCA in that their activities can be used as a measure of interaction of two target proteins. There are many different enzymes considered for PCA, including dihydrofolate reductase (DHFR), β-lactamase, β-galactosidase, aminoglycoside phosphotransferase, and hygromycin B phosphotransferase.[32,33]

Being the first enzyme used for PCA, DHFR is especially applied to the PCAs dealing with survival selection of libraries because this enzyme is essential for cell survival in both prokaryotes and eukaryotes. Two approaches are used to detect DHFR activity responses in the PCAs: one is to monitor cell survival in DHFR-negative cells grown in the absence of nucleotides, and the other is to use DHFR inhibitor fluorescein-methotrexate, which binds to reconstituted DHFR with high affinity, analyzing the binding with fluorescence systems.[33]

Because two split β-galactosidase fragments rapidly combine to form active enzyme in solution, in a PCA strategy, β-galactosidase is often split into donor and acceptor; the donor is conjugated with a fusion protein so that addition of a molecule that can bind the donor-fusion protein interrupts the donor-acceptor complementation.[32] Hence this complementation strategy is used to screen unknown target libraries based on the competition between the donor-fusion protein and unknown target molecules such as G protein–coupled receptor (GPCR) agonists or protease inhibitors.[34,35]

Unlike other enzymes used for PCA, β-lactamase is solely from prokaryotes and is not present in eukaryotes. Hence the β-lactamase PCA has no background in eukaryotic cells. In addition, β-lactamase is a small monomeric protein whose crystal structure is known,[36] allowing for rational design of protein fragments. In β-lactamase PCA, two split fragments of β-lactamase do not spontaneously interact with each other until they are brought into proximity by virtue of interaction of two conjugated proteins. The complementation response of two β-lactamase fragments can be visualized as well as quantified by fluorescence methods including fluorescence microscopy and flow cytometry.

6.1.4 SCOPE OF THE CHAPTER

The technique in which β-lactamase can be split into two fragments that can complement each other when they are properly presented was first shown by others.[37,38] We first introduced the use of this technique for type I transmembrane protein interactions such as TLR-TLR as well as TLR-adaptor interactions.[17] This chapter deals with the β-lactamase complementation techniques to detect in vivo interaction of TLRs and signaling adaptors as well as TLR-TLR dimerization based on our previous study.

6.2 THE IN VIVO β-LACTAMASE COMPLEMENTATION ASSAY STRATEGY

Based on the β-lactamase structure[36] as well as previous β-lactamase complementation studies,[37–39] two split fragments, *Bla(a)* and *Bla(b)*, have been determined, which consist of amino acids 26–196 and 198–290, respectively.[17] Each fragment is conjugated to TLR4 or MyD88 by a flexible linker $(Gly_4Ser)_3$ (Figure 6.2A). To assure that spontaneous fragment complementation does not occur, we use noninteracting proteins as a control, such as TLR4 and MyD88 death domain, which do not provide a PCA response (Figure 6.2A).

The β-lactamase substrate CCF2/AM is a cell-permeable fluorescent molecule.[40] As illustrated in Figure 6.2B, this molecule diffuses across the cell membrane and the cytoplasmic esterases hydrolyze its ester functionalities (Ac: acetyl; Bt: butyryl; and AM: acetoxymethyl), releasing the β-lactamase substrate CCF2. Excitation of the coumarin donor in CCF2 at 409 nm leads to fluorescence resonance energy transfer (FRET) to the fluorescein acceptor–generating emission of green fluorescence at 520 nm. If the TLR-TLR or TLR-MyD88 interactions bring two β-lactamase fragments into proximity, resulting in a complete form of β-lactamase, CCF2 is further hydrolyzed by β-lactamase, separating donor and acceptor. This separation leads to FRET disruption, and the isolated coumarin donor emits blue fluorescence at 447 nm (Figure 6.2B).

6.2.1 CLONING OF THE β-LACTAMASE FRAGMENTS, *BLA(A)* AND *BLA(B)*

Two β-lactamase gene fragments, *Bla(a)* and *Bla(b)*, can be amplified by PCR using any cloning vector coding the ampicillin resistance gene as a PCR template. The PCR template used in this study is pQE30 expression vector (Qiagen, Valencia, CA). The DNA polymerase used in this study is PfuUltra HF DNA polymerase (Stratagene, La Jolla, CA). The PCR product of the *Bla(a)* or *Bla(b)* gene sequence is tagged with a *Bam*HI site followed by a flexible linker, $(Gly_4Ser)_3$, on the 5′-end, and with an *Eco*RV site on the 3′-end.

**Protocol 6.1: PCR Cloning of Two Split
β-Lactamase Fragments, *Bla(a)* and *Bla(b)***

1. PCR primers for *Bla(a)* and *Bla(b)* gene amplification are as follows (the restriction sites are bold):
   Bla(a)-forward: 5′-GAA GAA **GGA TCC** GGA GGA GGA GGA AGT GGA GGA GGA GGA AGT GGA GGA GGA GGA AGT CAC CCT GAA ACG CTG GTG AAA GTA A-3′
   Bla(a)-reverse: 5′-GAA TTC **GAT ATC** TCA GCC AGT TAA TAG TTT GCG CAA CGT TGT-3′
   Bla(b)-forward: 5′-GTC GAC **GGA TCC** GGT GGT GGT GGT AGT GGT GGT GGT GGT AGT GGT GGT GGT GGT AGT CTA CTT ACT CTA GCT TCT CGG CAA CAA-3′
   Bla(b)-reverse: 5′-GAA TTC **GAT ATC** TTA CCA ATG CTT AAT CAG TGA GGC ACC TAT-3′
2. PCR reaction mixtures:
   10 ng/µl template (pQE30 vector): 1 µl
   20 µM forward primer: 1 µl

(a)

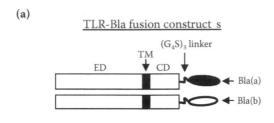

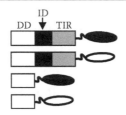

(b)

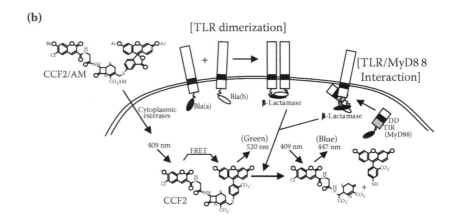

FIGURE 6.2   The In Vivo β-Lactamase Complementation Assay

Note: (A) β-lactamase fragment fusion constructs. (B) Schematic representation of the in vivo TLR binding assay using β-lactamase PCA. ED: extracellular domain; TM: transmembrane; CD: cytoplasmic domain; DD: death domain; ID: intermediate domain; TIR: Toll/IL-1R homology domain; and
FRET: fluorescence resonance energy transfer.

Source: Modified after Lee, H-K., Dunzendorfer, S., and Tobias, P. S. Cytoplasmic domain-mediated
dimerizations of Toll-like receptor 4 observed by β-lactamase enzyme fragment complementation. *J Biol Chem*, 279, 10564, 2004.

20 μM reverse primer: 1 μl
10 mM dNTPs: 1 μl
10× PfuUltra HF reaction buffer: 1 μl
2.5 U/μl PfuUltra HF DNA polymerase: 1 μl
Sterile water: 44 μl
Total reaction volume: 50 μl

3. PCR conditions:
   Segment 1 (1 cycle): 95 °C for 2 minutes
   Segment 2 (30 cycles): 95 °C for 30 seconds, 62 °C for 30 seconds, and 72 °C
       for 1 minute
   Segment 3 (1 cycle): 72 °C for 10 minutes, followed by 4 °C ending
4. Analyze the PCR product on a 1% agarose gel at a constant voltage of 85 V
   for 45 minutes.
5. Confirm that the bands are the expected size by examining the gel using
   UV.
6. Cut out the bands using a scalpel, and put them in microfuge tubes.
7. Purify the PCR products using the Gel Extraction Kit (Cat. 28704, Qiagen,
   Valencia, CA) according to the manufacturer's instructions.
8. Digest the PCR products with *Bam*HI and *Eco*RV.
9. Run the digested DNA on a 1% agarose gel, and purify the bands as described
   above (steps 4 through 7).
10. Store the purified DNA at −20 °C until needed (for up to several years).
11. Additionally, a full-length *Bla* gene is prepared by the procedures described
    above except for using the primer set of *Bla(a)*-forward and *Bla(b)*-reverse
    for PCR.

6.2.2   PCR Subcloning of TLR4, MyD88, and DD[MyD88ΔTIR] without a Stop Codon

The two expression constructs, pFLAG-CMV1/TLR4 expressing the wild-type
human TLR4 and pCDNA3.1 (+)/HA-MyD88 expressing the wild-type human
MyD88, are present in our laboratory (Figure 6.3). The TLR4 or MyD88 construct
without a stop codon as well as the DD[MyD88ΔTIR] construct are made based on
those original expression constructs. The PCR is designed to amplify a 1,509-base
pair (bp) TLR4 fragment (*Hpa*I-*Bam*HI) and an 891-bp MyD88 (*Hind*III-*Bam*HI;
Figure 6.3). Each PCR product contains these restriction enzyme sites: 5'-*Hpa*I and
3'-*Bam*HI for TLR4 fragments, and 5'-*Hind*III and 3'-*Bam*HI for MyD88 fragments.
Hence, each product is used to replace each corresponding fragment in the wild-type
constructs (Figure 6.3), resulting in a new construct lacking a stop codon.

Protocol 6.2: Subcloning of TLR4 and MyD88 without a Stop Codon

1. PCR primers are as follows (the restriction sites are bold):
   TLR4-forward: 5'-TGG CAA CAT TTA GAA TTA **GTT AAC** TGT AAA
       TTT GG-3'

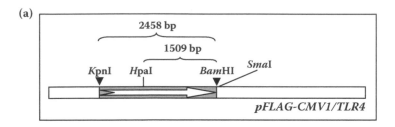

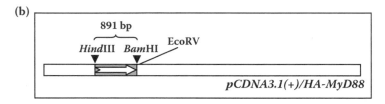

FIGURE 6.3   The Expression Vectors Including the Wild-Type-*TLR4* or -*MyD88* Gene
Note: pFLAG-CMV1/TLR4 contains a 2,458-bp *TLR4* gene between *Kpn*I and *Bam*HI sites (A), while
    pCDNA3.1(+)/HA-MyD88 contains a 891-bp *MyD88* gene between *Hind*III and *Bam*HI sites (B).

> TLR4-reverse: 5′-ATA ATA **GGA TCC** GAT AGA TGT TGC TTC CTG
> CC-3′
> MyD88-forward: 5′-ATA ATA **AAG CTT** ATG TAT CCT TAT GAT GTT
> CCT GAT TAT GCT GCT GC-3′
> MyD88-reverse: 5′-ATA ATA **GGA TCC** GGG CAG GGA CAA GGC CTT
> GGC-3′

2. Prepare the PCR mixtures as described in Protocol 6.1 (step 2) using pFLAG-CMV1/TLR4 and pCDNA3.1 (+)/HA-MyD88 (Figure 6.3) as the PCR templates for TLR4 PCR and MyD88 PCR, respectively.
3. PCR conditions:
   Segment 1 (1 cycle): 95 °C for 2 minutes
   Segment 2 (30 cycles): 95 °C for 30 seconds, 62 °C for 30 seconds, and
       72 °C for 1 minute (except TLR4 PCR, for which the last step is 72 °C
       for 2 minutes)
   Segment 3 (1 cycle): 72 °C for 10 minutes, followed by 4 °C ending
4. Analyze the PCR products, and purify as described in Protocol 6.1 (steps 4–7).
5. Digest the PCR products as well as expression vectors (Figure 6.3) using restriction enzymes as follows:
   *Hpa*I + *Bam*HI: TLR4 PCR product and pFLAG-CMV1/TLR4 vector
   *Hind*III + *Bam*HI: MyD88 PCR product and pCDNA3.1(+) vector (Cat. V790-20, Invitrogen, Carlsbad, CA)
6. Purify the digested DNA samples (two PCR fragments and two vector fragments) as described in Protocol 6.1 (step 9).
7. Ligate the purified DNAs using T4 DNA ligase (Cat. M0202S, New England Biolabs, Beverly, MA) at room temperature for 30 minutes. The ligation mixtures are as follows:

A. TLR4 fragment ligation:
    Sterile water: 11 µl
    PCR product/*Hpa*I+*Bam*HI (insert): 3 µl
    pFLAG-CMV1/TLR4/*Hpa*I+*Bam*HI (vector): 1 µl
    5× ligation buffer: 4 µl
    T4 DNA ligase: 1 µl
    Total volume: 20 µl
B. MyD88 fragment ligation:
    Sterile water: 11 µl
    PCR product/*Hind*III+*Bam*HI (insert): 3 µl
    pCDNA3.1(+)/*Hind*III+*Bam*HI (vector): 1 µl
    5× ligation buffer: 4 µl
    T4 DNA ligase: 1 µl
    Total volume: 20 µl

8. Add 3 µl of the ligation mixture to competent DH5α cells (Cat. 18265-017, Invitrogen, Carlsbad, CA), and incubate for 30 minutes on ice.
9. Heat the cells at 37 °C for 30 seconds, and leave on ice for 2 minutes.
10. Mix the cells with 1 ml of SOC medium (Cat. 15544-034, Invitrogen), and shake at 37 °C for 1 hour.
11. Spread the cells on LB agar containing 100 µg/ml ampicillin, and incubate overnight at 37 °C.
12. Perform colony PCR for finding a colony with the right insert.
13. Select 10 colonies, and mix each colony with 50 µl sterile water in a microfuge tube.
14. Heat at 95 °C for 5 minutes, centrifuge at 12,000 rpm for 2 minutes, and use the supernatants as the PCR templates.
15. Mix the following:
    Template: 5 µl
    20 µM forward primer (as described above): 1 µl
    20 µM reverse primer (as described above): 1 µl
    10 mM dNTPs: 1 µl
    10× PCR buffer (Cat. 18067-017, Invitrogen): 1 µl
    25 mM MgCl$_2$ (Cat. 18067-017, Invitrogen): 1 µl
    5 U/µl Taq DNA polymerase (Cat. 18038, Invitrogen): 0.5 µl
    Sterile water: 39.5 µl
    Total reaction volume: 50 µl
PCR conditions:
    Segment 1 (1 cycle): 94 °C for 3 minutes
    Segment 2 (30 cycles): 94 °C for 30 seconds, 62 °C for 30 seconds, and 72 °C for 3 minutes
    Segment 3 (1 cycle): 72 °C for 10 minutes, followed by 4 °C ending
16. Analyze the PCR products, and purify as described in Protocol 6.1 (steps 4–5).
17. Choose two right colonies per construct, and culture in 3 ml LB/Amp (100 µg/ml) broth overnight.

18. Purify the plasmid DNAs using the Qiagen Spin Miniprep Kit (Cat. 27104, Qiagen, Valencia, CA).
19. Determine the DNA concentrations by measuring the optical density at 260 nm, and store at –20 °C until needed.

Protocol 6.3: Subcloning of DD[MyD88ΔTIR] without a Stop Codon

1. PCR primers are as follows (the restriction sites are bold):
   DD-forward: ATT ATA **AAG CTT** ATG TAT CCT TAT GAT GTT CCT GAT TAT GCT TCC CTT CCC CTG GCT GCT CTC AAC ATG CG
   DD-reverse: ATA ATA **GGA TCC** AAT GCT GGG TCC CAG CTC CAG CAG
2. Prepare the PCR mixtures as described in Protocol 6.1 (step 2).
3. Follow the PCR conditions as described in Protocol 6.2 (step 3).
4. Analyze the PCR products and purify as described in Protocol 6.1 (steps 4–7).
5. Digest the PCR products using *Hind*III and *Bam*HI, and purify as described in Protocol 6.1 (step 9).
6. Ligate the digested PCR product with the pCDNA3.1(+) vector cut with *Hind*III+*Bam*HI in Protocol 6.2.
7. Mix the following:
   Sterile water: 11 µl
   PCR product/*Hind*III+*Bam*HI (insert): 3 µl
   pCDNA3.1(+)/*Hind*III+*Bam*HI (vector): 1 µl
   5× ligation buffer: 4 µl
   T4 DNA ligase: 1 µl
   Total volume: 20 µl
8. Perform DH5α cell transformation, select the colonies containing recombinant DNA, and purify the plasmid DNA as described in Protocol 6.2 (steps 8–19).

6.2.3   SUBCLONING OF β-LACTAMASE FRAGMENT FUSION CONSTRUCTS

Protocol 6.4: Subcloning of TLR4/*Bla(a)* and TLR4/*Bla(b)* Constructs

1. Digest pFLAG-CMV1/TLR4ΔStop Codon made in Protocol 6.2 with *Bam*HI and *Sma*I.
2. Purify the digested DNA as described in Protocol 6.1 (step 9).
3. Ligate the purified DNA with the *Bla(a)* or *Bla(b)* fragment prepared in Protocol 6.1. Ligation mixtures:

Sterile water: 11 µl

Bla(a)/*Bam*HI+*Eco*RV or *Bla(a)*/*Bam*HI+*Eco*RV (insert): 3 µl

pFLAG-CMV1/TLR4ΔStop Codon/*Bam*HI+*Sma*I (vector): 1 µl

5× ligation buffer: 4 µl

T4 DNA ligase: 1 µl

Total volume: 20 µl

4. Perform DH5α cell transformation and colony PCR as described in Protocol 6.2 (steps 8–16).

5. Mix the following:

Template: 5 µl

20 µM TLR4-forward: 1 µl

20 µM *Bla(a)*-reverse (or *Bla[b]*-reverse): 1 µl

10 mM dNTPs: 1 µl

10× PCR buffer (Invitrogen, Carlsbad, CA): 1 µl

25 mM $MgCl_2$ (Invitrogen): 1 µl

5 U/µl Taq DNA polymerase (Invitrogen): 0.5 µl

Sterile water: 39.5 µl

Total reaction volume: 50 µl

PCR conditions:

Segment 1 (1 cycle): 94 °C for 3 minutes

Segment 2 (30 cycles): 94 °C for 30 seconds, 62 °C for 30 seconds, and 72 °C for 3 minutes

Segment 3 (1 cycle): 72 °C for 10 minutes, followed by 4 °C ending

6. Choose two right colonies per construct and purify DNA as described in Protocol 6.2 (steps 17–19).

7. Confirm that no PCR errors occur by DNA sequencing.

8. Purify the DNA using the EndoFree Plasmid Maxi Kit (Cat. 12362, Qiagen, Valencia, CA), determine the DNA concentrations by measuring the optical density at 260 nm, and store at –20 °C until needed.

Protocol 6.5: Subcloning of MyD88/*Bla(a)* and MyD88/*Bla(b)* Constructs

1. Digest pCDNA3.1(+)/MyD88ΔStop Codon made in Protocol 6.2 with *Bam*HI and *Eco*RV.

2. Purify the digested DNA as described in Protocol 6.1 (step 9).

3. Ligate the purified DNA with the *Bla(a)* or *Bla(b)* fragment prepared in Protocol 6.1. Ligation mixtures:

Sterile water: 11 µl

Bla(a)/*Bam*HI+*Eco*RV or *Bla(a)*/*Bam*HI+*Eco*RV (insert): 3 µl

pCDNA3.1(+)/MyD88ΔStop Codon/*Bam*HI+*Eco*RV (vector): 1 µl

5× ligation buffer: 4 µl

T4 DNA ligase: 1 µl

Total volume: 20 µl

4. Perform DH5α cell transformation and colony PCR as described in Protocol 6.2 (steps 8–16).
5. Mix the following:
>   Template: 5 μl
>   20 μM MyD88-forward: 1 μl
>   20 μM *Bla(a)*-reverse (or *Bla[b]*-reverse): 1 μl
>   10 mM dNTPs: 1 μl
>   10× PCR buffer (Invitrogen, Carlsbad, CA): 1 μl
>   25 mM MgCl$_2$ (Invitrogen): 1 μl
>   5 U/μl Taq DNA polymerase (Invitrogen): 0.5 μl
>   Sterile water: 39.5 μl
>   Total reaction volume: 50 μl

   PCR conditions:
>   Segment 1 (1 cycle): 94 °C for 3 minutes
>   Segment 2 (30 cycles): 94 °C for 30 seconds, 62 °C for 30 seconds, and 72 °C for 3 minutes
>   Segment 3 (1 cycle): 72 °C for 10 minutes, followed by 4 °C ending

6. Choose two right colonies per construct and purify DNA as described in Protocol 6.2 (17–19).
7. Confirm that no PCR errors occur by DNA sequencing.
8. Purify the DNA using the EndoFree Plasmid Maxi Kit (Cat. 12362; Qiagen, Valencia, CA), determine the DNA concentrations by measuring the optical density at 260 nm, and store at –20 °C until needed.
9. Additionally, a β-lactamase-positive construct of MyD88/full-length *Bla* is made using the full-length *Bla* gene described in Protocol 6.1 (step 11) by repeating this protocol (6.5).

Protocol 6.6: Subcloning of DD/*Bla(a)* and DD/*Bla(b)* Constructs

1. Digest pCDNA3.1(+)/DD[MyD88ΔTIR] made in Protocol 6.3 with *Bam*HI and *Eco*RV.
2. Purify the digested DNA as described in Protocol 6.1 (step 9).
3. Ligate the purified DNA with the *Bla(a)* or *Bla(b)* fragment prepared in Protocol 6.1. Ligation mixtures:
>   Sterile water: 11 μl
>   *Bla(a)*/*Bam*HI+*Eco*RV or *Bla(a)*/*Bam*HI+*Eco*RV (insert): 3 μl
>   pCDNA3.1(+)/DD[MyD88ΔTIR]/*Bam*HI+*Eco*RV (vector): 1 μl
>   5× ligation buffer: 4 μl
>   T4 DNA ligase: 1 μl
>   Total volume: 20 μl

4. Perform DH5α cell transformation and colony PCR as described in Protocol 6.2 (steps 8–16).
5. Mix the following:

Template: 5 μl
20 μM DD-forward: 1 μl
20 μM *Bla(a)*-reverse (or *Bla[b]*-reverse): 1 μl
10 mM dNTPs: 1 μl
10× PCR buffer (Invitrogen, Carlsbad, CA): 1 μl
25 mM $MgCl_2$ (Invitrogen): 1 μl
5 U/μl Taq DNA polymerase (Invitrogen): 0.5 μl
Sterile water: 39.5 μl
Total reaction volume: 50 μl

PCR conditions:
Segment 1 (1 cycle): 94 °C for 3 minutes
Segment 2 (30 cycles): 94 °C for 30 seconds, 62 °C for 30 seconds, and 72 °C for 3 minutes
Segment 3 (1 cycle): 72 °C for 10 minutes, followed by 4 °C ending

6. Choose two right colonies per construct and purify DNA as described in Protocol 6.2 (17–19).
7. Confirm that no PCR errors occur by DNA sequencing.
8. Purify the DNA using the EndoFree Plasmid Maxi Kit (Cat. 12362, Qiagen, Valencia, CA), determine the DNA concentrations by measuring the DNA optical density at 260 nm, and store at –20 °C until needed.

6.3   THE β-LACTAMASE PCAs

The β-lactamase PCAs described in this chapter consist of fluorescence microscopy and flow cytometry. Each assay procedure deals with mammalian cell culture and transient transfection of β-lactamase fragment chimeras. Two major cell lines used are human embryonic kidney 293 (HEK293) cells and human epithelial cervical carcinoma (HeLa) cells.

6.3.1   PREPARATION OF HEK293 CELLS OR HELA
CELLS FOR TRANSIENT TRANSFECTION

HEK293 cells or HeLa cells are grown in Dulbecco's modified Eagle's medium (high-glucose DMEM; Cat. 10313, Invitrogen, Carlsbad, CA) supplemented with 10% fetal bovine serum (Cat. SH30070.03, HyClone, Logan, UT) and 1% penicillin-streptomycin-glutamine (100×, prepared with 10,000 U/ml penicillin G sodium, 10 mg/ml streptomycin sulfate, 29.2 mg/ml L-glutamine, and 10 mM sodium citrate in 0.14% NaCl; Cat. 10378-016, Invitrogen). Cells are cultured in a 75 cm^2 tissue culture flask for routine growth, in a 94% air and 6% CO_2 mixture, at 37 °C.

Protocol 6.7: Culture of HEK293 Cells and HeLa Cells

1. Remove all culture medium from a 75 cm^2 flask with cells grown for 2–3 days.

2. Add 3 ml of 0.05% Trypsin-EDTA (Cat. 25300, Invitrogen, Carlsbad, CA), and incubate at 37 °C for 5 minutes.
3. Detach cells from the surface of the flask by gently rocking, and add 17 ml of warmed fresh medium.
4. Transfer 2 ml of cell suspension to a new 75 cm^2 flask containing 18 ml of warmed fresh medium, and incubate at 37 °C for 2–3 days.
5. Continue the routine cell culture with each cell line in duplicate while the PCAs are conducted.

6.3.2   COMPLEMENTATION ANALYSIS BY FLUORESCENCE MICROSCOPY

Protocol 6.8: Preparation of HEK293 Cells or HeLa Cells for Transient Transfection

1. Remove all culture medium from a 75 cm^2 flask with cells grown for 2–3 days.
2. Add 3 ml of 0.05% Trypsin-EDTA, and incubate at 37 °C for 5 minutes.
3. Detach cells from the surface of the flask by gently rocking, and add 17 ml of warmed fresh medium.
4. Transfer all cell suspension in a 50 ml tube, and centrifuge at 1,000 rpm for 5 minutes.
5. Remove all existing medium from the tube, resuspend cells with 10 ml of fresh medium, and count cells in a hemocytometer.
6. Plate cells (1×10^6 cells/ml/well) in 12-well tissue culture plates, and incubate at 37 °C overnight.

Protocol 6.9: Transient Transfection of HEK293 Cells or HeLa Cells

1. DNA constructs and co-transfection sets:
   A. Vector control (pFLAG-CMV1 or pCDNA3.1[+] vector)
   B. TLR4/*Bla(a)* + TLR4/*Bla(b)*
   C. MyD88/*Bla(a)* + MyD88/*Bla(b)*
   D. DD[MyD88]/*Bla(a)* + MyD88/*Bla(b)*
   E. MyD88/*Bla(a)* + TLR4/*Bla(b)*
   F. DD[MyD88]/*Bla(a)* + TLR4/*Bla(b)*
   G. MyD88/full *Bla*
2. Prepare transfection mixtures A and B separately as follows:
   A. DNA mixture (per well):
      DNA: 0.25 µg
      Opti-Mem I (Cat. 31985, Invitrogen, Carlsbad, CA): 50 µl

     B.Lipofectamine mixture (per well):

        Lipofectamine 2000 (Cat. 11668-027, Invitrogen): 2 μl

        Opti-Mem I: 50 μl

3. Mix the two mixtures after 5 minutes, and incubate at room temperature for 20 minutes.

4. Add the mixture to each well of the 12-well plates containing HEK293 or HeLa cells prepared in Protocol 6.8.

5. Incubate the cell plates at 37 °C overnight.

Protocol 6.10: CCF2/AM Substrate Treatment

1. Remove all existing medium from the cell plates containing the transfected cells in Protocol 6.8, and add 1 ml of physiological saline buffer (10 mM HEPES, 6 mM sucrose, 10 mM glucose, 140 mM NaCl, 5 mM KCl, and 2 mM $CaCl_2$ [pH 7.35]) to each well.

2. Prepare the substrate CCF2/AM solution as follows:

    1 mM CCF2/AM (Cat. K1032 loading kit, Invitrogen, Carlsbad, CA) in DMSO: 1 μl

    Solution B (Cat. K1032 loading kit, Invitrogen): 9 μl

    250 mM Probenecid (Cat. P-8761, Sigma-Aldrich, St. Louis, MO) in 250 mM NaOH: 10 μl

    Physiological saline buffer: 1 ml

    Total volume (per well): 1 ml

3. Remove physiological saline buffer from the cell plates, add the substrate solution, and incubate at room temperature for 1 hour.

4. Remove the substrate solution and add 1 ml of physiological saline buffer containing 2.5 mM Probenecid.

5. Proceed to fluorescence microscopy (Protocol 6.11) or flow cytometry (Protocols 6.12 and 6.13).

Protocol 6.11: Complementation Analysis by Fluorescence Microscopy

1. The equipment used in this study is as follows:

    A. Fluorescence microscope (Zeiss Axiovert 100 TV, Zeiss, Göttingen, Germany) with a Diagnostics Instruments SPOT cooled CCD camera (Diagnostics Instruments, Sterling Heights, MI)

    B. Filter set (Omega Optical XF12-2, Omega Optical, Brattleboro, VT):

        Excitation filter: XF1007 (405DF10)

        Dichroic mirror: XF2005 (420DCLP)

        Emission filter: XF3088 (435ALP)

2. Open SPOT program from a computer connected with the SPOT cooled CCD camera.
3. Place a cell plate, which is prepared in Protocol 6.10, on the microscope stage, and examine the cells with 10× and 32× objective lenses.
4. Capture the cell images by the SPOT CCD camera using the SPOT program according to the manufacturer's instructions. The captured images are shown in Figure 6.4.

6.3.3   COMPLEMENTATION ANALYSIS BY FLOW CYTOMETRY

HEK293 or HeLa cells are transient-transfected with the β-lactamase fusion constructs and treated with CCF2/AM substrate as described in Protocols 6.8 and 6.9. These cells are further prepared for flow cytometry analysis (Figure 6.5).

Protocol 6.12: Preparation of the Substrate-Treated Transfectants for Flow Cytometry

1. Remove all existing substrate solution from the cell plates in Protocol 6.10 (step 4).
2. Add 0.5 ml of the Enzyme-Free, PBS-Based Cell Dissociation Buffer (Cat. 13151-014, Invitrogen, Carlsbad, CA) to each well, and incubate at 37 °C for 10 minutes.
3. Add 1.5 ml of basic sorting buffer (1× phosphate buffered saline [Ca/Mg^{++} free], 1% fetal bovine serum, 1 mM EDTA, and 25 mM HEPES [pH 7.0]), which is 0.2 μm–filter sterilized.
4. Transfer the cell suspensions in each well to a 15 ml tube separately, and centrifuge at 1,000 rpm for 5 minutes.
5. Remove the supernatants, and resuspend the cell pellet with 0.5 ml basic sorting buffer.
6. Transfer the cell suspensions to FACS tubes (Cat. 352052, BD Biosciences, Franklin Lakes, NJ).

Protocol 6.13: Flow Cytometry Analysis

1. The flow cytometry analysis in this study is performed using the BD LSR II equipped with band pass filters of 440/40 nm and 525/50 nm, which is combined with BD FACSDiva Software (BD Biosciences, Franklin Lakes, NJ), and all the assays follow the FACSDiva manual provided by the manufacturer.
2. Set up your experiment in a hierarchical format: Folder → Experiment → Specimen → Tube.

(a)

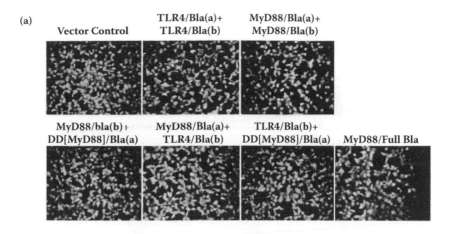

(b)

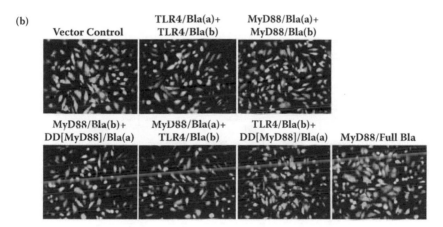

FIGURE 6.4 (Color figure follows page 104) In Vivo β-Lactamase Complementation Analyzed by Fluorescence Microscopy

Note: HEK293 cells (A) or HeLa cells (B) were transiently co-transfected with different combinations of the β-lactamase fusion constructs (as noted) and loaded with CCF2/AM.

3. Open "Tube_001" and select the instrument parameters, including Alexa Fluor 430 (AF430), AF405, Forward Scatter (FSC), and Side Scatter (SSC), in the "Instrument Window."

4. Indicate the labels of Violet 525 for AF430 and Violet 440 for AF405 in the "Inspector" panel, and set up the threshold values in the "Instrument Window."

5. Create several dot plots, including SSC-A versus FSC-A, FSC-W versus FSC-A, SSC-W versus SSC-A, and AF405-A versus AF430-A, on the Instrument Window.

6. Run a control sample without recording by clicking "Acquire," and control the voltage to properly locate the cell population in the plots.

(b)

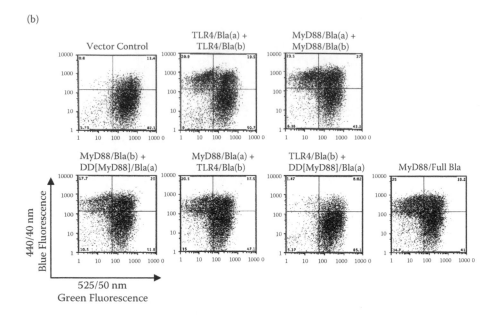

(b)

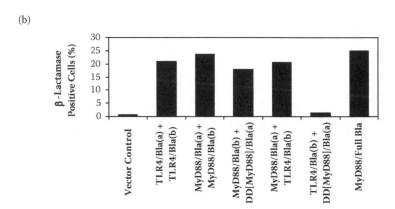

FIGURE 6.5   In Vivo β-Lactamase Complementation Analyzed by Flow Cytometry

Note: HEK293 cells were transiently co-transfected with different combinations of the β-lactamase fusion constructs (as noted) and loaded with CCF2/AM. The cells were then harvested and analyzed by FACS. The β-lactamase-positive cells were observed in the upper-left quadrant (A), and the percentage population of β-lactamase-positive cells was plotted (B).

7. Start "Compensation" and sequentially gate the proper population of cells. The appropriate conditions in this study are as follows:

Voltage: 279 (FSC), 275 (SSC), 300 (AF430), and 400 (AF405)

Compensation: 8 (AF430) and 21 (AF405)

8. Start running the samples and collecting the data by clicking "Record" for each sample.

9. Save the collected data, and proceed to data analysis.

10. As shown in Figure 6.5A, each dot plot is divided into quadrants, in which the upper-left quadrant represents percentage complementation. The complementation rates can be analyzed using the Excel program, as shown in Figure 6.5B.

REFERENCES

1. Janeway, C. A., Jr., and Medzhitov, R. Introduction: the role of innate immunity in the adaptive immune response. *Semin Immunol*, 10, 349, 1998.
2. Aderem, A., and Ulevitch, R. J. Toll-like receptors in the induction of the innate immune response. *Nature*, 406, 782, 2000.
3. Akira, S., Uematsu, S., and Takeuchi, O. Pathogen recognition and innate immunity. *Cell*, 124, 783, 2006.
4. Ozato, K., Tsujimura, H., and Tamura, T. Toll-like receptor signaling and regulation of cytokine gene expression in the immune system. *BioTech*, 33, S66, 2002.
5. Tosi, M. F. Innate immune response to infection. *J Allergy Clin Immunol*, 116, 241, 2005.
6. Trinchieri, G., and Sher, A. Cooperation of Toll-like receptor signals in innate immune defence. *Nature Rev Immunol*, 7, 179, 2007.
7. Medzhitov, R. Toll-like receptors and innate immunity. *Nat Rev Immunol*, 1, 135, 2001.
8. Akira, S., Takeda, K., and Kaisho, T. Toll-like receptors: critical proteins linking innate and acquired immunity. *Nat Rev Immunol*, 2, 675, 2001.
9. Hoebe, K., Janssen, E., and Beutler, B. The interface between innate and adaptive immunity. *Nat Immunol*, 5, 971, 2004.
10. Choe, J., Kelker, M. S., and Wilson, I. A. Crystal structure of human toll-like receptor 3 (TLR3) ectodomain. *Science*, 309, 581, 2005.
11. Bell, J. K., Botos, I., Hall, P. R., Askins, J., Shiloach, J., and Segal, D. M. The molecular structure of the Toll-like receptor 3 ligand-binding domain. *Proc Natl Acad Sci USA*, 102, 10976, 2005.
12. Ozinsky, A., Underhill, D. M., Fontenot, J. D., Hajjar, A. M., Smith, K. D., Wilson, C. B., Schroeder, L., and Aderem, A. The repertoire for pattern recognition of pathogens by the innate immune system is defined by cooperation between Toll-like receptors. *Proc Natl Acad Sci USA*, 97, 13766, 2000.
13. Hajjar, A. M., O'Mahony, D. S., Ozinsky, A., Underhill, D. M., Aderem, A., Klebanoff, S. J., and Wilson, C. B. Cutting edge: functional interactions between toll-like receptor (TLR) 2 and TLR1 or TLR6 in response to phenol-soluble modulin. *J Immunol*, 166, 15, 2001.
14. Zhang, H., Tay, P. N., Cao, W., Li, W., and Lu, J. Integrin-nucleated Toll-like receptor (TLR) dimerization reveals subcellular targeting of TLRs and distinct mechanisms of TLR4 activation and signaling. *FEBS Lett*, 532, 171, 2002.
15. Tapping, R. I., and Tobias, P. S. Mycobacterial lipoarabinomannan mediates physical interactions between TLR1 and TLR2 to induce signaling. *J Endotoxin Res*, 9, 264, 2003.
16. Underhill, D. M. Toll-like receptors: networking for success. *Eur J Immunol*, 33, 1767, 2003.
17. Lee, H-K., Dunzendorfer, S., and Tobias, P. S. Cytoplasmic domain-mediated dimerizations of Toll-like receptor 4 observed by β-lactamase enzyme fragment complementation. *J Biol Chem*, 279, 10564, 2004.

18. Buchanan, S. G., and Gay, N. J. Structural and functional diversity in the leucine-rich repeat family of proteins. *Pro Biophys Mol Biol*, 65, 1, 1996.
19. Girardin, S. E., Sansonetti, P. J., and Philpott, D. J. Intracellular vs extracellular recognition of pathogens: common concepts in mammals and flies. *Trends Microbiol*, 10, 193, 2002.
20. Medzhitov, R., Preston-Hurlburt, P., and Janeway, C. A., Jr. A human homologue of the *Drosophila* Toll protein signals activation of adaptive immunity. *Nature*, 388, 394, 1997.
21. Ulevitch, R. J. Therapeutics targeting the innate immune system. *Nat Rev Immunol*, 4, 512, 2004.
22. Wagner, H. Endogenous TLR ligands and autoimmunity. *Adv Immunol*, 91, 159, 2006.
23. Heil, F., Ahmad-Nejad, P., Hemmi, H., Hochrein, H., Ampenberger, F., Gellert, T., Dietrich, H., Lipford, G., Takeda, K., Akira, S., Wagner, H., and Bauer, S. The Toll-like receptor 7 (TLR7)-specific stimulus loxoribine uncovers a strong relationship within the TLR7, 8 and 9 subfamily. *Eur J Immunol*, 33, 2987, 2003.
24. Latz, E., Schoenemeyer, A., Visintin, A., Fitzgerald, K. A., Monks, B. G., Knetter, C. F., Lien, E., Nilsen, N. J., Espevik, T., and Golenbock, D. T. TLR9 signals after translocating from the ER to CpG DNA in the lysosome. *Nat Immunol*, 5, 190, 2004.
25. Lee, H-K., Dunzendorfer, S., Soldau, K., and Tobias, P. S. Double-stranded RNA-mediated TLR3 activation is enhanced by CD14. *Immunity*, 24, 153, 2006.
26. Kawai, T., Adachi, O., Ogawa, T., Takeda, K., and Akira, S. Unresponsiveness of MyD88-deficient mice to endotoxin. *Immunity*, 11, 115, 1999.
27. Ishii, K. J., Coban, C., and Akira, S. Manifold mechanisms of Toll-like receptor-ligand recognition. *J Clin Immunol*, 25, 511, 2005.
28. O'Neill, L. A., and Bowie, A. G. The family of five: TIR-domain-containing adaptors in Toll-like receptor signaling. *Nat Rev Immunol*, 7, 353, 2007.
29. Yamamoto, M., Sato, S., Hemmi, H., Sanjo, H., Uematsu, S., Kaisho, T., Hoshino, K., Takeuchi, O., Kobayashi, M., Fujita, T., Takeda, K., and Akira, S. Essential role for TIRAP in activation of the signaling cascade shared by TLR2 and TLR4. *Nature*, 420, 324, 2002.
30. Horng, T., Barton, G. M., Flavell, R. A., and Medzhitov, R. The adaptor molecule TIRAP provides signaling specificity for Toll-like receptors. *Nature*, 420, 329, 2002.
31. Yamamoto, M., Sato, S., Hemmi, H., Uematsu, S., Hoshino, K., Kaisho, T., Takeuchi, O., Takeda, K., and Akira, S. TRAM is specifically involved in the Toll-like receptor 4-mediated MyD88-independent signaling pathway. *Nat Immunol*, 4, 1144, 2003.
32. Eglen, R. M. Enzyme fragment complementation: a flexible high throughput screening assay technology. *Assay Drug Dev Technol*, 1, 97, 2002.
33. Remy, I., and Michnick, S. W. Mapping biochemical networks with protein-fragment complementation assays. *Methods Mol Biol*, 261, 411, 2004.
34. Golla, R., and Seethala, R. A homogeneous enzyme fragment complementation cyclic AMP screen for GPCR agonists. *J Biomol Screen*, 7, 515, 2002.
35. Naqvi, T., Lim, A., Rouhani, R., Singh, R., and Eglen, R. M. β-galactosidase enzyme fragment complementation as a high-throughput screening protease technology. *J Biomol Screen*, 9, 398, 2004.
36. Strynadka, N. C. J., Adachi, H., Jensen, S. E., Johns, K., Sielecki, A., Betzel, C., Sutoh, K., and James, M. N. G. Molecular structure of the acyl-enzyme intermediate in β-lactam hydrolysis at 1.7 Å resolution. *Nature*, 359, 700, 1992.
37. Galarneau, A., Primeau, M., Trudeau, L. E., and Michnick, S. W. β-lactamase protein fragment complementation assays as *in vivo* and *in vitro* sensors of protein-protein interactions. *Nat Biotechnol*, 20, 619, 2002.

38. Wehrman, T., Kleaveland, B., Her, J. H., Balint, R. F., and Blau, H. M. Protein-protein interactions monitored in mammalian cells via complementation of β-lactamase enzyme fragments. *Proc Natl Acad Sci USA*, 99, 3469, 2002.
39. Spotts, J. M., Dolmetsch, R. E., and Greenberg, M. E. Time-lapse imaging of a dynamic phosphorylation-dependent protein-protein interaction in mammalian cells. *Proc Natl Acad Sci USA*, 99, 15142, 2002.
40. Zlokarnik, G., Negulescu, P. A., Knapp, T. E., Mere, L., Burres, N., Feng, L., Whitney, M., Roemer, K., and Tsien, R. Y. Quantitation of transcription and clonal selection of single living cells with beta-lactamase as reporter. *Science*, 279, 84, 1998.

7 Use of Toll-Like Receptor Chimeras to Dissect Mechanisms of Receptor Localization and Signaling

Tadashi Nishiya and Anthony L. DeFranco

CONTENTS

7.1 BACKGROUND

Immune protection in mammals results from the combination of innate immune recognition molecules that provide genome-encoded sensing of conserved components of infectious organisms[1] with the somatic recombination-generated antigen recognition molecules of lymphocytes that recognize individualized features of particular infecting organisms. Among the molecules that provide hardwired recognition of infection by bacteria, fungi, parasites, and viruses, the Toll-like receptors (TLRs) are perhaps the most important for immune function. They are present in the three types of immune cells resident in tissues—the macrophage, the immature dendritic cell, and the mast cell—all of which act as sentinels that respond to recognition of infection by TLRs by inducing inflammation and/or taking antigen to the draining lymph node to initiate an adaptive immune response. TLRs are also present in a variety of other cell types, including B lymphocytes, endothelial cells, and epithelial cells, the latter two of which may also contribute to initial recognition of infection in some circumstances.

TLRs are composed of a ligand-binding N-terminal extracellular domain made up of leucine-rich repeats that pack together to form a horseshoe structure,[2] a single hydrophobic transmembrane domain, and a C-t erminal cytoplasmic domain made up of a short spacer followed by a globular TIR domain, named for its presence in *Drosophila melanogaster* Toll and in the mammalian TLRs, in the interleukin-1 receptor, and in many plant disease-resistance gene products. The TIR domain is also present in five intracellular adaptor molecules in mammals, and these adaptors interact with subsets of TLRs via TIR-TIR domain interactions to mediate their signaling.[3]

Whereas the TLRs that recognize bacterial cell wall components are present on the surface of macrophages and other cells, the TLRs that recognize nucleic acid ligands are localized to intracellular compartments.[4] Both types of TLRs are thought to recognize their ligands within phagosomes, but presumably those TLRs present on the cell surface can also recognize their ligands prior to phagocytosis. Therefore, it has been proposed that the intracellular localization of TLR3, TLR7, TLR8, and TLR9 is an adaptation to minimize responses to the nucleic acids of the host.[5] Interestingly, in this regard, there is accumulating evidence that TLR7 and TLR9 participate in the aberrant production of the anti-DNA, antichromatin, and antiribonucleoprotein autoantibodies seen in most patients with the autoimmune disease systemic lupus erythematosus.[6] Clearly there is a lot still to be learned with regard to how the nucleic acid–sensing TLRs distinguish self–nucleic acids, such as those present in an apoptotic cell that has been phagocytosed by a macrophage or a dendritic cell, from the viral or bacterial nucleic acids present in an apoptotic virus-infected cell or in phagocytosed virions or microbes, but the restriction of these TLRs to intracellular compartments is likely to be part of the answer.

The function of TLRs has been elucidated by three main approaches: (1) definition of the ligands for individual TLRs coupled with analysis of the responses of immune cells to those ligands; (2) genetic loss-of-function experiments, especially analysis of mice with natural or targeted ablation of particular TLRs; and (3) the use of chimeric TLRs to dissect the signaling properties and signaling mechanisms of this family of receptors. With regard to the chimeric TLR approach, chimeras have been constructed that constitutively dimerize or trimerize TLR TIR domains, and such chimeras induce constitutive signaling,[7–10] suggesting that ligand-induced oligomerization may initiate TLR signaling. The constitutive nature of these chimeras, however, makes it difficult to study the mechanisms of receptor signaling, and, moreover, the high-level expression of these molecules in nonimmune cells may not fully reflect functional properties of TLRs in immune cells. Therefore, we set out to develop a TLR chimera system to dissect the functionally distinct properties of individual TLRs in the context of ligand-induced oligomerization and expression at roughly normal levels in immune cells. For this purpose, we made chimeras of TLRs in which the extracellular domain was derived from TLR4 and the transmembrane domain and intracellular domain were derived from another TLR. These chimeric molecules were expressed in bone marrow–derived macrophages (BMDMs) at roughly normal levels by a highly efficient viral transduction method. In addition, the macrophages used were derived from C57BL/10ScN mice, which carry a deletion of the TLR4 gene,[11–14] so when LPS was added to these cells, it would oligomerize the chimeric TLRs via binding to the extracellular domain of TLR4 complexed with MD-2, but would not bind any endogenous TLRs.

7.1.1 Intracellular Localization of Nucleic Acid–Recognizing TLRs

The TLR chimeras made using the extracellular domain of TLR4 exhibited a cellular localization that was very similar to that of the TLR that supplied the transmembrane domain and cytoplasmic domain. Chimeras with TLR1, TLR2, TLR5, and TLR6 were expressed on the cell surface at levels similar to that of full-length TLR4 expressed from the same retroviral vector.[15] In contrast, the chimera with TLR8 was expressed on the cell surface at a low but detectable level, and the chimeras with TLR3, TLR7, and TLR9 were expressed inside the cell but not on the cell surface. Subsequent studies demonstrated that these chimeras co-localized strongly with full-length TLR3 or TLR7 fused to green fluorescent protein (GFP) or one of its derivatives.[16] These TLRs were found to localize to abundant small vesicles within macrophages. This compartment was not definitively identified, although it was shown to be different from early endosomes. Moreover, TLR3 and TLR7 were seen to accumulate in or near phagosomes containing apoptotic T cells, suggesting that these TLRs are able to sense ligands within phagosomes. Other studies have suggested that TLR9 localizes to the endoplasmic reticulum.[17] In any case, this localization was specified by the transmembrane and/or cytoplasmic domains of these receptors.

The mechanism by which TLR3 and TLR7 are retained within intracellular compartments was explored by a random mutagenesis approach. The transmembrane domain– and cytoplasmic domain–encoding regions of the cDNA were subjected to random mutagenesis by error-prone PCR. The mutagenized sequences were reinserted into a vector containing the extracellular domain of TLR4 and enhanced GFP downstream from an internal ribosome entry site (IRES) sequence, and pools of approximately 40,000 mutagenized TLRs were transfected into HEK293T cells that stably expressed transfected CD14 and MD-2. Intact transfected cells were stained with a fluorescent anti-TLR4/MD-2 antibody, and cells positive for both GFP and TLR4/MD-2 were isolated by cell sorting, which was repeated a second time to further enrich for mutated versions of the TLR chimera that had lost intracellular localization.[16]

Mutations of TLR3 causing loss of intracellular retention were mostly frameshift mutations that truncated the cytoplasmic domain of TLR3 within the spacer sequence between the transmembrane domain and the TIR domain. In contrast, deletion of just the TIR domain did not change the localization of the TLR3 chimera. These studies identified a 23-amino-acid sequence as being necessary for intracellular localization of the TLR4/TLR3 chimera. Grafting this membrane proximal spacer sequence onto the heterologous protein CD25 (IL-2 receptor α chain) caused CD25 to adopt a primarily intracellular localization, demonstrating that this sequence is also largely sufficient to mediate internal localization. This sequence is not similar to any of the well-characterized sequences known to target intracellular membrane proteins to their location, suggesting that TLR3 uses a novel mechanism to achieve its localization.

TLR3 was found to co-localize strongly with TLR7,[16] but surprisingly, there was no clear homology between the membrane proximal spacer region of TLR3 and the corresponding region in TLR7. Moreover, random mutagenesis of the TLR4/TLR7 chimera did not yield any mutants that relocalized the TLR4/TLR7 chimera to the cell surface. Subsequent site-directed mutagenesis of the TLR4/TLR7 chimera demonstrated that the membrane proximal spacer region and the TIR domain were both dispensable for intracellular location and that the transmembrane domain was responsible.[16] Remarkably, TLR3 and TLR7 use different regions of their protein sequence to target themselves to the identical intracellular localization in macrophages.

7.1.2 DISTINCTIVE SIGNALING PROPERTIES OF TLRs

The TLR chimeras that were expressed exclusively in intracellular vesicles failed to respond to LPS,[16] suggesting that LPS does not easily gain access to this compartment, in agreement with direct demonstration that these chimeras did not localize to early endosomes.

The TLR chimeras that are expressed on the cell surface mediated responses to LPS, including production of TNF and activation of signaling responses, such as NF-κB activation and activation of the Erk, JNK, and p38 MAP kinases.[15] These responses were seen in cells expressing full-length TLR4 and in cells expressing the TLR4/TLR5 chimera, indicating that the cytoplasmic domains of TLR5 were able to signal efficiently in the context of the TLR4/TLR5 chimera. Whereas the TLR4/TLR5 chimera was reasonably efficient at inducing TNF production and signaling events downstream of MyD88,[15] it was only weakly capable of inducing type 1 interferon production and subsequent STAT1 tyrosine phosphorylation. The latter responses were strongly induced by LPS stimulation of cells expressing full-length TLR4, presumably because the cytoplasmic domain of TLR4 associates with the adaptor TRIF, which can induce interferon-β production, whereas the cytoplasmic domain of TLR5 does not. The TLR4/TLR8 chimera was able to reduce responses similar to those induced by the TLR4/TLR5 chimera, except at a lower level, presumably reflecting the low level of cell surface expression of this chimera.

In contrast, chimeras containing the transmembrane and cytoplasmic domains of TLR1, TLR2, or TLR6 were expressed on the cell surface at levels similar to those of full-length TLR4 or the TLR4/TLR5 chimera, but failed to induce TNF or signaling responses to LPS when expressed alone in bone marrow–derived macrophages. In contrast, in macrophages expressing TLR4/TLR2 together with either TLR4/TLR1 or TLR4/TLR6, LPS induced robust production of TNF, substantial activation of NF-κB, and activation of MAP kinases.[16] In none of these cases was the combination of TLR chimeras able to activate a substantial amount of interferon-β production. Thus, the cytoplasmic domains of TLR2, TLR1, and TLR6, all of which have TIR domains, exhibit functional complementarity in certain combinations (TLR2 with TLR1 or TLR6), but not in other combinations (TLR1 with TLR6), and activate the MyD88 signaling pathway but not the TRIF signaling pathway.

Previous work had indicated that TLR2 functions together with TLR1 for responses to triacylated lipoproteins and that TLR2 functions together with TLR6 for responses to diacylated lipoproteins.[1] These distinct lipoproteins are found in bacterial cell walls. Previously, it had been thought that the complementarity of TLR2 with TLR1 or TLR6 reflects binding specificity of the extracellular domains of TLR1 and TLR6, and this is likely true, but our findings indicate that in addition TLR2 requires a complementary TLR TIR domain to promote efficient signaling. Interesting in this regard is that additional ligands for TLR2 have been identified, including peptidoglycan, porins from *Neisseria*, lipoarabinomannan from mycobacteria, and phospholipomannan from certain fungal pathogens. For these ligands, a second TLR in addition to TLR2 is likely to be required for signaling to be generated, but this complementary TLR has not yet been identified.

7.2 CONSTRUCTION OF TLR CHIMERAS

All TLRs are type I transmembrane proteins composed of an NH_2-terminal signal peptide, an extracellular domain involved in ligand recognition, a single transmembrane domain, and a cytoplasmic domain largely made up of the TIR domain. All TLRs have a cysteine-rich region proximal to the transmembrane domain in their extracellular region. Four cysteine residues are located in this region. Mutation analysis of *Drosophila* Toll protein has shown that these cysteine residues are important for its function.[18] For this reason, the highly conserved last cysteine in the cysteine-rich region of TLR4 and other TLRs was used to align the sequences and to provide the junctions of the chimeras (Figure 7.1). We used two distinct methods to construct TLR chimeras: a triple ligation method and the PCR sewing method, as described in Figure 7.2.

7.2.1 TRIPLE LIGATION METHOD

This method is simple. As shown in Figure 7.2A, cDNAs encoding the extracellular region of TLR4 (4EX) and the transmembrane and cytoplasmic regions of other TLRs (TM-CP) were amplified by PCR using primers containing desired restriction enzyme sites and 5′ phosphate modification: the 5′ primer for 4EX and the 3′ primer for TM-CP contained a restriction enzyme site, and the 3′ primer for 4EX and the 5′

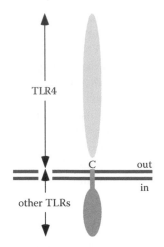

FIGURE 7.1   The TLR Chimeras

Note: TLR chimeras are composed of the extracellular region of
TLR4 and the transmembrane and cytoplasmic regions of
other TLRs. The highly conserved last cysteine (c) in the
cysteine-rich regions of TLR4 and other TLRs is used to
align the sequences and to provide the junctions of the
chimera.

primer for TM-CP had a 5′ phosphate modification, which is indispensable for the
blunt-end ligation between 4EX and TM-CP. The PCR products were digested with
restriction enzymes and then ligated with the vector.

Protocol 7.1: Construction of TLR Chimeras by a Triple Ligation Method

1. Amplify cDNA encoding 4EX and TM-CP of other TLRs by PCR using
   PfuUltra high-fidelity DNA polymerase (Stratagene, Santa Clara, CA). The
   composition of the reaction is below.

   distilled H_2O: 78 µl
   10× buffer: 10µl
   DMSO: 5µl
   0.1 µg/µl template: 1µl
   200 ng/µl upper primer: 1.25 µl
   200 ng/µl lower primer: 1.25 µl
   10 mM dNTP: 2 µl
   PfuUltra: 1.5 µl
   Total volume: 100 µl (50 µl × 2 tubes)

   The PCR is performed for 18 cycles at 95 °C for 1 minute, 55 °C for 1 minute, and 72 °C for 3 minutes and 30 seconds. After the last cycle, the
   reaction is incubated at 72 °C for 10 minutes to complete extension of
   PCR products. Eighteen cycles is enough to amplify PCR products;
   additional cycles would increase the chance of PCR errors.
2. Purify PCR products by using the Gel Purification Kit (Qiagen, Valencia,
   CA). Gel purification is used to confirm that PCR products are successfully
   amplified.
3. Digest PCR products with restriction enzymes overnight.
4. Purify PCR products by using the PCR Purification Kit (Qiagen).

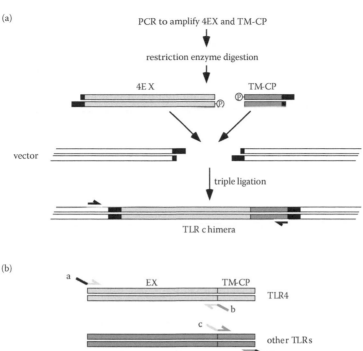

FIGURE 7.2 Two Methods Used for Construction of TLR Chimeras

Note: (A) Triple ligation method. cDNAs encoding the extracellular region of TLR4 (4EX) and the transmembrane and cytoplasmic regions of other TLRs (TM-CP) were amplified by PCR. After restriction enzyme digestion, the triple ligation was performed. The primers shown for the final product
were used to screen transfected bacterial colonies. (B) PCR sewing method. The cDNAs encoding
4EX and TM-CP were amplified in the first PCR reaction. The second PCR was performed by
using primers a and d and the first PCR products as templates. The resulting PCR products encode
TLR chimeras, which were then digested with restriction enzymes and ligated with the vector. The
gray bar shows 4EX, the dark-gray bar shows TM-CP, the open bar shows vector sequence, and
black bars show restriction enzyme sites. Primers a and d contain a restriction enzyme site. Primers
b and c contain the other TLR sequence and the TLR4 sequence, respectively, so that an overlap-
ping sequence can be created between the two PCR products.

5. Ligate 4EX, TM-CP, and vector. The molecular ratio of 4EX, TM-CP, and vector should be 10:1:1. This ratio was chosen to optimize the efficiency of the triple ligation because TM-CP (about 600 bp) is much smaller than 4EX (about 1,900 bp). In general, the ligation efficiency varies inversely in proportion to the size of the fragment.

6. Transform *E. coli* competent cells with ligation mixture.

7. Perform colony PCR to identify bacterial colonies carrying vectors incorporating the TLR chimeras. A vector sequence upstream of the insert was used for the upper primer, and the lower primer used to amplify TM-CP was used for the lower primer of colony PCR (Figure 7.2A). The composition of this reaction is as follows:

Distilled H_2O: 91.7 µl
DMSO: 10.9 µl
200 ng/µl upper primer: 2.2 µl
200 ng/µl lower primer: 2.2 µl
GoTaq Green Master Mix (Promega, Madison, WI): 109 µl
Total volume: 218 µl (for 10 colonies; 20 µl × 10 tubes)

This PCR is performed for 24 cycles at 95 °C for 1 minute, 55 °C for 1 minute, and 72 °C for 3 minutes.

8. DNA sequencing to confirm that the chimera's sequence is correct.

The triple ligation method is simple, but the ligation efficiency is not high; usually, 60% of the bacterial colonies carried vectors containing TM-CP alone, about 20% of the bacterial colonies carried vectors containing TLR chimeras, and the rest of the colonies carried self-ligated vectors. The ligation of vector with TM-CP alone and the self-ligation of vector were not prevented, although use of different restriction enzyme sites for 4EX and TM-CP to the vector helped to minimize unwanted products. Therefore, to find bacterial colonies carrying plasmids containing TLR chimeras, at least 16 different colonies were screened by colony PCR.

7.2.2   PCR Sewing Method

This method is composed of two steps of PCR. As described in Figure 7.2B, in the first step of PCR, cDNA encoding 4EX containing a part of TM-CP was amplified using primers a and b. Similarly, cDNA encoding TM-CP of other TLRs containing a part of 4EX was amplified using primers c and d. Primers a and d also contained restriction enzyme sites, whereas b and c contained the sequence encoding a part of 4EX or TM-CP of other TLRs. In the second step, the cDNA encoding the TLR chimera was amplified by using primers a and d and PCR products obtained from the first-step PCR as templates. The second PCR products encode TLR chimeras, which are then digested with restriction enzymes to generate sticky ends for ligation into the vector.

Protocol 7.2: Construction of TLR Chimeras by a PCR Sewing Method

1. Perform first PCR to amplify cDNAs encoding 4EX and TM-CP by using PfuUltra high-fidelity DNA polymerase (Stratagene, Santa Clara, CA), as described in Protocol 2.1.
2. Perform second PCR to obtain cDNA encoding TLR chimeras. The composition of this reaction is as follows:

   10× buffer: 10 µl
   DMSO: 5µl
   4EX: 30 ng
   TM-CP: 10 ng
   200 ng/µl upper primer: 1.25 µl
   200 ng/µl lower primer: 1.25 µl
   10 mM dNTP: 2 µl
   PfuUltra: 1.5 µl
   Add distilled H_2O to 100 µl.

   This PCR is performed for 18 cycles at 95 °C for 1 minute, 55 °C for 1 minute, and 72 °C for 3 minutes and 30 seconds. After the last cycle, the reaction is incubated at 72 °C for 10 minutes to complete extension of PCR products. The molecular ratio of 4EX to TM-CP is 1:1.
3. Purify PCR products by using the Gel Purification Kit (Qiagen, Valencia, CA).
4. Digest PCR products with restriction enzymes overnight.
5. Purify PCR products by using the PCR Purification Kit (Qiagen).
6. Ligate PCR products with vector. The molecular ratio of PCR product to vector is 10:1.
7. Transform *E. coli* competent cells with ligation mixture.
8. Perform colony PCR to identify colonies carrying TLR chimeras, as described in Protocol 7.1.
9. Sequence DNA to confirm correct sequence.

The ligation efficiency of the PCR sewing method is much higher than that of the triple ligation method. Although this method has more steps than the triple ligation method, most of the bacterial colonies carried the vector with the sequences encoding the TLR chimeras.

7.3 INTRODUCTION OF TLR CHIMERAS INTO BONE MARROW–DERIVED MACROPHAGES BY RETROVIRAL GENE TRANSFER

Because the major functions of TLRs are to induce innate immune responses in immune cells, such as macrophages and dendritic cells, assessing the function of the TLR chimeras in those cells was desired. However, innate immune cells are difficult to transfect efficiently by using typical transfection methods, such as liposome-based methods, DEAE-dextran, or calcium-phosphate. In contrast, VSV-G pseudotype retroviruses efficiently infect primary macrophages. In this protocol, we describe a retroviral gene transfer to introduce TLR chimeras into bone marrow–derived macrophages with excellent efficiency.

7.3.1 Preparation of Bone Marrow–Derived Macrophages from Mice

Typically, we obtained 2–3×10^7 macrophages from one mouse (two sets of femora and tibias).

Protocol 7.3: Preparation of Bone Marrow–Derived Macrophages from Mice

1. Euthanize the mouse with carbon dioxide gas followed by cervical dislocation.
2. Sterilize the mouse fur with 70% ethanol.
3. Remove the skin, fur, and muscles. Next, cut the bone that joins the spine and patella.
4. Clean the bones with Kimwipes (Kimberly-Clark, Dallas, TX) as much as possible. Then, wash the exterior of the bones with 70% ethanol, and store in tissue culture medium until the cleanup of the last bone is finished. *Note*: Subsequent steps are performed in a laminar flow biosafety cabinet.
5. Holding the bones with sterile forceps, cut off each end using sterile scissors.
6. Flush the bones with approximately 2.5 ml of medium per femur or tibia, using a 26 gauge needle or syringe at both ends until the bones turn white.
7. Centrifuge the collected marrow at $500 \times g$ for 5 minutes at 4 °C. The pellets should be red because of red blood cells.
8. Resuspend the pellets with 5 ml of red blood cell lysis buffer, and incubate for 5 minutes at room temperature.
9. Add 5 ml of medium to the cells, and centrifuge. After centrifugation, the cell pellet should be white or light pink.
10. Resuspend the cells in 20 ml of macrophage medium. We use CMG12-14 cell culture supernatant for CSF-1 source.[19] Alternatively, L929 cell supernatant can be used at 30% of final concentration.
11. Spread the bone marrow cells onto a 15 cm plastic Petri dish (not treated for tissue culture). *Note*: The compositions of the red blood cell lysis buffer and the primary macrophage media are described below.

Red Blood Cell Lysis Buffer (500 ml)

Hepes: 1.19 g (10 mM)
NH_4Cl: 4.1 g (150 mM)
$KHCO_3$ 0.5 g (10 mM)
0.5 mM EDTA: 100 μl (0.1 mM)

Add water to 500 ml. Adjust pH to 7.3 with 1 M KOH. Sterilize through 0.2 μm filter, and store at room temperature.

Culture Media for Bone Marrow–Derived Macrophages (500 ml)

RPMI 1640 containing 25 mM Hepes: 400 ml
FBS: 50 ml
100 mM sodium pyruvate: 5 ml
CMG12-14 cell condition medium: 45 ml

7.3.2 PREPARATION OF RETROVIRUSES

VSV-G retroviruses producing TLR chimeras are prepared by triple transfection of HEK293T cells with retroviral constructs along with gag-pol and VSV-G expression constructs.[20]

Protocol 7.4: Preparation of Retroviruses

1. Split HEK 293T cells at a density of 1.8×10^6 cells per well of a 6-well plate.
2. The next day, transform the cells with 3.6 μg of retroviral vector, 1.4 μg of gag-pol expression vector, and 2.1 μg of VSV-G expression vector by using 10 μl of LipofectAMINE 2000 reagent (Invitrogen, Carlsbad, CA).
3. Twenty-four hours after transfection, change medium with fresh media (DMEM containing 10% FBS), and further incubate the cells for 24 hours.
4. Collect the media containing viruses (virus supernatants), and use these media for infection or store them at −70 °C until use.

VSV-G retroviruses are stable for at least 1 year at −70 °C, and the titer is not significantly changed by a couple of freeze-thaw steps. Virus supernatants can be collected three times; repeat steps 3 and 4 (above) three times.

7.3.3 INFECTION OF BONE MARROW–DERIVED MACROPHAGES WITH RETROVIRUSES

Approximately 30–50% of bone marrow–derived macrophages can be successfully infected with VSV-G retroviruses by this protocol.

Protocol 7.5: Infection of Bone Marrow–Derived Macrophages with Retroviruses

1. Plate BMDMs at day 4 or 5 after the preparation from the bone marrow at a density of 7.5×10^5 cells per well in a nontissue culture–treated 6-well plate.

2. The next day, mix viral supernatants with polybrene at a final concentration of 10 μg/ml, and then add to the cells.

3. Centrifuge cells and viruses at 1,300 × g for 1 hour at room temperature. This is a critical step to obtain maximum infection efficiency.

4. Incubate the cells for 6 hours at 37 °C in 5% CO_2 and 95% air. In general, 6 hours of incubation is enough to complete the integration of virus into the genome of cells.

5. Replace viral supernatants with culture media. After 48 hours, assess the infection efficiency by flow cytometric analysis as described in the next protocol.

It is critical to use BMDMs at day 4 to 6 after the preparation from the bone marrow for retroviral infection, because the infection efficiency is significantly decreased at day 7 or later. The efficiency of retroviral infection is totally dependent on the proliferation rate of cells. BMDMs proliferate very fast until day 6, and suddenly become slow after day 7.

Before mixing with polybrene, the virus supernatants should be centrifuged at 500 × g for 10 minutes at 4 °C to remove contaminating HEK293T cells. The filtration of viral supernatants with a 0.45 μm filter to remove HEK293T cells from viral supernatants is not recommended because the titer of retroviruses is significantly decreased by this operation.

It has been suggested that the concentration of retroviruses by polyethylene glycol (PEG) precipitation increases virus titer. However, we have observed no significant increment of virus titer by PEG precipitation.

7.3.4 DETERMINATION OF INFECTION EFFICIENCY BY FLOW CYTOMETRY

Retroviral vectors carrying an IRES element and the enhanced green fluorescent protein (eGFP) gene are useful to easily assess the infection efficiency for cells. We have used the bicistronic retroviral vector pMX-pie[21] for the introduction of TLR chimeras into BMDMs. By using this vector, the cells expressing the gene of interest also express eGFP proteins. Therefore, eGFP fluorescence is an indication of infection efficiency and expression level of the gene of interest. eGFP fluorescence is simply detected by the use of a flow cytometer. In this protocol, we describe the preparation of cells for flow cytometry.

Protocol 7.6: Determination of Infection Efficiency by Flow Cytometry

1. Perform retroviral infection of BMDMs as described in Protocol 7.5.

2. Forty-eight hours after infection, incubate the cells in phosphate buffered saline (PBS) containing 5 mM EDTA for 5–10 minutes at 37 °C.

3. Detach the cells from the plate by gentle pipetting. Cells should be easily detached if plastic plates that have not been treated for tissue culture are used.

4. Centrifuge the cells at 900 × g for 1 minute at 4 °C.

TABLE 7.1

The Efficiency of Infection in Bone Marrow–Derived Macrophages

TLR Chimeras	% Green Fluorescent Protein + Cells
TLR4/TLR1	36.8 ± 6.4
TLR4/TLR2	37.5 ± 5.6
TLR4/TLR3	36.3 ± 4.1
TLR4	37.0 ± 6.4
TLR4/TLR5	38.7 ± 5.6
TLR4/TLR6	38.0 ± 4.6
TLR4/TLR7	34.5 ± 1.3
TLR4/TLR8	36.7 ± 4.7
TLR4/TLR9	35.0 ± 3.2
Vector	52.3 ± 4.3

Note: Infection efficiency of TLR chimeras was assessed as described in Protocol 7.6, and was shown as average ± *SD* of three independent infections.

5. Resuspend the cells with FACS-staining buffer consisting of 1% FBS and 0.09% sodium azide in PBS.
6. Measure GFP fluorescence by flow cytometer. As shown in Table 7.1, approximately 35% of BMDMs express GFP in our experiments.

7.4 CHARACTERIZATION OF TLRS BY USING TLR CHIMERAS

TLR chimeras are useful to examine both subcellular localization and signaling properties of TLRs. In this protocol, we describe protocols for studies of subcellular localization and signaling of TLRs by using TLR chimeras.

7.4.1 EXAMINATION OF THE SUBCELLULAR LOCALIZATION OF TLRS BY FLOW CYTOMETRY

Flow cytometry is a useful method to determine whether the receptor is expressed on the cell surface or resides in intracellular compartments. To detect receptors on the cell surface, intact cells are stained with antibodies specific to the extracellular portion of the receptors of interest. To detect receptors inside cells, fixation and permeabilization of the cells are needed before staining with antibodies. Miyake's group has generated an anti-mouse TLR4/MD-2 antibody,[22] which recognizes the extracellular region of TLR4 associated with MD-2, an accessory molecule required for cell surface expression and LPS responsiveness of TLR4.[23] Because the TLR chimeras constructed by the method described above possess the extracellular region of

TLR4 (Figure 7.1), the anti-mouse TLR4/MD-2 antibody is a useful tool to examine the subcellular distribution of the TLR chimeras.

To specifically detect TLR chimeras with the anti-mouse TLR4/MD-2 antibody, TLR chimeras must be expressed in macrophages prepared from TLR4-deficient mice. We have prepared bone marrow–derived macrophages from C57BL/10ScN mice, which carry a null mutation in the TLR4 gene.[11,12,14] In this protocol, we describe the sample preparation for flow cytometry using an anti-mouse TLR4/MD-2 antibody.

Protocol 7.7: Measuring the Expression Levels of TLR Chimeras on the Cell Surface and inside the Cells by Flow Cytometry

1. Prepare BMDMs from TLR4-deficient mice as described in Protocol 7.3.
2. Infect the cells at day 4 or 5 with retroviruses producing TLR chimeras as described in Protocol 7.5.
3. Forty-eight hours after infection, detach the cells by incubating in PBS containing 5 mM EDTA, followed by gentle pipetting.
4. Treat the cells with an anti-CD16/CD32 antibody (clone 2.4G2, BD Bioscience, San Jose, CA) to block Fcγ receptors II and III.
5. After washes, stain the cells with Phycoerythrin (PE) conjugated anti-mouse TLR4/MD-2 antibody (eBioscience, San Diego, CA) for 30 minutes at 4 °C. For intracellular staining, the cells are fixed and permeabilized with Cytofix/Cytoperm solution (BD Bioscience) before staining with PE–anti-mouse TLR4/MD-2 antibody.
6. After washes, resuspend the cells in FACS-staining buffer.
7. Measure the expression level of TLR chimeras on the cell surface by flow cytometry.

Figure 7.3 shows the subcellular distribution of TLR chimeras examined by flow cytometry. TLR4/TLR1, TLR4/TLR2, TLR4, TLR4/TLR5, and TLR4/TLR6 are expressed on the cell surface, whereas TLR4/TLR3, TLR4/TLR7, TLR4/TLR8, and TLR4/TLR9 are localized intracellularly. These data are consistent with data obtained by available anti-TLR antibodies or fluorescently labeled TLRs. The subcellular distribution of TLRs may be regulated by their transmembrane and/or cytoplasmic regions. It should be noted that if the extracellular region contains a targeting element, these TLR chimeras may not reflect the correct cellular localization of TLRs.

FIGURE 7.3 Subcellular Distribution of the TLR Chimeras Determined by Flow Cytometry

Note: Macrophages infected with retroviruses producing TLR chimeras were stained with anti-mouse TLR4/MD-2 antibody. GFP-positive cells were gated and displayed for TLR chimera staining. Intact: cells without fixation/permeabilization; fixed/permeabilized: cells with fixation/permeabilization: dotted line: isotype control; and solid line: anti-TLR4/MD-2 antibody.

Source: From Nishiya, T., and DeFranco, A. L. Ligand-regulated chimeric receptor approach reveals distinctive subcellular localization and signaling properties of the Toll-like receptors. *J. Biol. Chem.* 279, 19008, 2004. With permission.

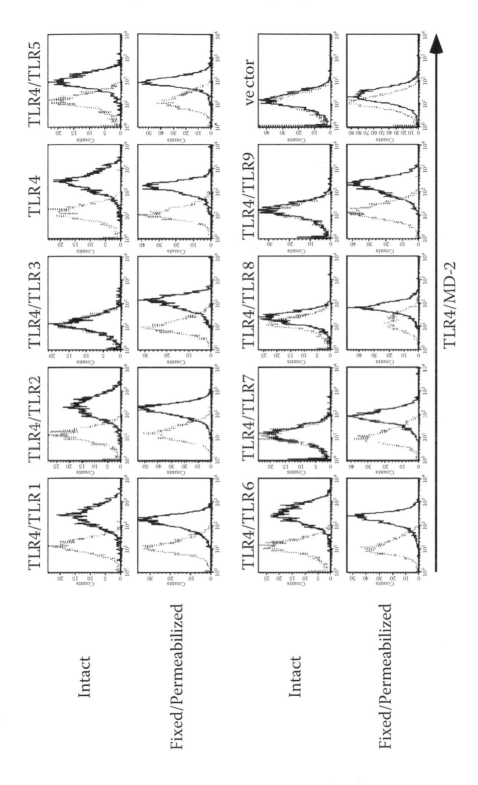

7.4.2 EXAMINATION OF THE SUBCELLULAR
LOCALIZATION OF TLRs BY MICROSCOPY

Fluorescent microscopy is useful to characterize the intracellular organelle, where intracellular receptors are localized. Anti-mouse TLR4/MD-2 antibody is suitable for microscopy as well as flow cytometry. Because antibodies against some TLRs have not been established yet, the microscopy of TLR chimeras using anti-mouse TLR4/MD-2 antibody has given us important information about TLR localization in cells.

Protocol 7.8: Determination of the Subcellular
Localization of TLR Chimeras by Microscopy

1. Prepare BMDMs from TLR4-deficient mice as described in Protocol 7.3.
2. Infect the cells at day 4 or 5 with retroviruses producing TLR chimeras as described in Protocol 7.5.
3. Plate the cells at a density of 9×10^5 cells per dish in a 35 mm glass-bottom dish.
4. Next day, treat the cells with anti-CD16/CD32 antibody to block Fcγ receptors II and III.
5. Wash the cells with 2 ml of FACS-staining buffer three times.
6. Fix and permeabilize the cells with 100 µl of Cytofix/Cytoperm solution for 20 minutes at 4 °C.
7. Wash the cells with 2 ml of Perm Wash buffer (BD Bioscience, San Jose, CA) three times.
8. Treat the cells with appropriate antibodies against various subcellular marker proteins.
9. Wash the cells as in step 7.
10. Treat the cells with PE–anti-mouse TLR4/MD-2 antibody for 40 minutes at 4 °C.
11. Wash the cells as in step 7.
12. Treat the cells with 300 nM DAPI for 5 minutes at room temperature to visualize nucleus.
13. After washes with FACS-staining buffer, add 1 ml of FACS-staining buffer to the dish and perform microscopy. PE signals bleach very rapidly. You can stain the cells with unlabeled anti-mouse TLR4/MD-2 antibody and fluorescently-labeled anti-rat IgG antibody instead of PE-labeled anti-mouse TLR4/MD-2 antibody.

7.4.3 DETERMINATION OF THE INTRACELLULAR TARGETING
ELEMENTS OF TLRs BY RANDOM MUTAGENESIS APPROACH

Error-prone PCR-based random mutagenesis can be applied to the identification of the targeting elements for intracellular TLRs: TLR3, TLR7, TLR8, and TLR9. The subcellular distribution of TLRs appears to be regulated by their transmembrane and/ or cytoplasmic regions. The introduction of mutation(s) into the targeting elements may disrupt the conformational characteristics involved in the intracellular localization of TLRs, and therefore the mutants will be expressed on the cell surface.

There are some important considerations for performing this experiment. First, a retroviral vector should be used to express the mutant TLRs, because it is critical to stably express distinct mutants in as many cells as possible. Second, mutants should be introduced to HEK293T cells stably expressing mouse CD14 and MD-2 (designated as 293TCM cells), because (1) the infection efficiency of HEK293T cells (> 90%) is much higher than that of primary macrophages (~ 35%), (2) the HEK293T cells do not express endogenous human TLR4, and (3) HEK293T cells are easily cloned because they proliferate well. Finally, supercompetent *E. coli* cells, such as cDNA library grade, should be used for the transformation with vectors carrying TLR4/TLR chimeras with point mutations (TLR4/TLR^{Mu} chimeras).

Protocol 7.9: Determination of the Intracellular Targeting Elements of TLRs by Random Mutagenesis Approach

1. The entire transmembrane and cytoplasmic regions of TLR chimeras are amplified by error-prone PCR using the Genemorph II Random Mutagenesis Kit (Stratagene, Santa Clara, CA) (Figure 7.4). The composition of reaction is given below.

   Distilled H_2O: 38.7 µl
   10× buffer: 5 µl
   DMSO: 2.5 µl
   10 ng/ml template: 0.5 µl
   200 ng/ml upper primer: 0.63 µl
   200 ng/ml lower primer: 0.63 µl
   10 mM dNTP: 1 µl
   Mutazyme II: 1 µl
   Total volume: 50 µl

   PCR is performed for 30 cycles at 95 °C for 30 seconds, 55 °C for 30 seconds, and 72 °C for 1 minute. After the last cycle, the reaction mixture is incubated at 72 °C for 10 minutes to complete extension of PCR products. The error rate can be controlled by both amount of template and cycle number of PCR reaction. Following this protocol, the mutation rates of the TLR4/TLR3 and TLR4/TLR7 chimeras were 1.7 ± 0.3 and 2.9 ± 0.4 per clone at the nucleotide level and 1.1 ± 0.4 and 1.6 ± 0.3 per clone at the amino acid level, respectively.

2. Purify PCR products by using the Gel Purification Kit (Qiagen, Valencia, CA).

3. Digest PCR products with *SalI* and *NotI* restriction enzymes overnight.

4. Purify PCR products by using the PCR Purification Kit (Qiagen).

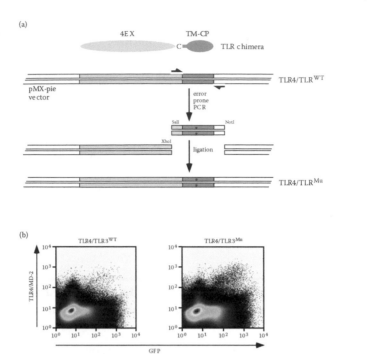

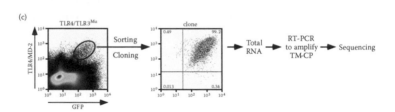

FIGURE 7.4 Random Mutagenesis of the Transmembrane and Cytoplasmic Regions of TLR Chimeras

Note: (A) Schematic of random mutagenesis. Random mutagenesis of the transmembrane and cytoplasmic regions of TLR chimeras was performed by error-prone PCR using primers indicated by the arrows. PCR products carrying mutation(s) were cloned into the pMX-pie-TLR4(EX) vector. (B) Flow cytometric analysis of cell surface expression of wild-type and mutated TLR4/TLR3 chimeras. HEK293TCM cells were infected with retroviruses encoding the TLR4/TLR3 chimera (TLR4/TLR3^{WT}) or encoding pooled chimeras containing random mutations in the transmembrane and cytoplasmic domains of TLR3 (TLR4/TLR3^{Mu}). Forty-eight hours after infection, the cell surface level of TLR4/TLR3 chimeras was determined by flow cytometric analysis using anti-mouse TLR4/MD-2 antibody. (C) Isolation of mutations in TLR4/TLR3^{Mu} chimeras that express on the cell surface. The GFP+ TLR4/MD-2+ population gated by the oval shown was sorted twice, and then individual cells were cloned by limiting dilution. About 200 clones were isolated, and the cell surface level of the TLR4/TLR3^{Mu} chimera in each clone was determined by flow cytometry using anti-mouse TLR4/MD-2 antibody.

Source: Adapted from Nishiya, T., Kajita, E., Miwa, S., and DeFranco, A. L. TLR3 and TLR7 are targeted to the same intracellular compartments by distinct regulatory elements. *J. Biol. Chem.* 280, 37107, 2005. With permission.

5. Ligate PCR products with the *XhoI/NotI* site of pMX-pie-TLR4(EX) vector, which contains the cDNA encoding amino acids 1–598 of TLR4.

6. Transform 200 µl of OmniMAX supercompetent *E. coli* cells (Invitrogen, Carlsbad, CA) with 20 µl of ligation mixture, and plate onto 5 × 15 cm LB agar plates. In the case of TLR4/TLR3^{Mu} and TLR4/TLR7^{Mu} chimera experiments, we obtained about 40,000 bacterial colonies from the plates.

7. Count and collect all colonies on the plates, and expand them in 200 ml of LB media.

8. Purify the plasmid DNAs from the pooled bacteria.

9. Generate a pool of retroviruses encoding TLR4/TLR^{Mu} chimeras.

10. Infect 293TCM cells with a pool of retroviruses encoding TLR4/TLR^{Mu} chimeras.

11. Twenty-four hours after infection, start puromycin selection.

12. Three days later, detect the cell surface TLR4/TLR3^{Mu} in puromycin-resistant cells by flow cytometry using PE-anti-mouse TLR4/MD-2 antibody.

13. Enrich the anti-mouse TLR4/MD-2 positive cells by using a cell sorter. Cells with TLR4/MD-2 expressed on the surface are obtained by two rounds of sorting to enrich strongly for real positive cells.

14. Clone the anti-mouse TLR4/MD-2 positive cells by limiting dilution; about 40 cells are resuspended with 20 ml media, distributed in the wells of 96-well plates (200 µl/well), and cultured for 2 weeks.

15. Test individual clones expressing mutated TLR4 chimeras for cell surface expression by flow cytometry.

16. Prepare total RNAs from each clone, and reverse-transcribe.

17. Amplify the resulting cDNA by PCR, and perform DNA sequencing to determine the nature of the mutation causing relocalization to the cell surface.

7.4.4 Examination of the Signaling Properties of TLRs

TLR chimeras are expressed in BMDMs from TLR4-deficient mice, and stimulated with a TLR4 ligand (Figure 7.5). Highly purified LPS is necessary for this experiment, because crude LPS preparation may contain ligands for other TLRs, such as peptidoglycan or lipopeptides, which may induce a response in untransfected CS7BL/10SCN macrophages. Highly purified LPS is now commercially available from several companies.

Protocol 7.10: Examination of the Signaling Properties of TLRs

1. Prepare BMDMs from TLR4-deficient mice as in Protocol 7.3.

2. Infect the cells on day 4 or 5 with retroviruses producing TLR chimeras as described in Protocol 7.5.

3. Plate the cells at a density of 4×10^5 cells per well in wells of a 24-well plate, or at a density of 9×10^5 cells per well of a 6-well plate. We usually use

(a)

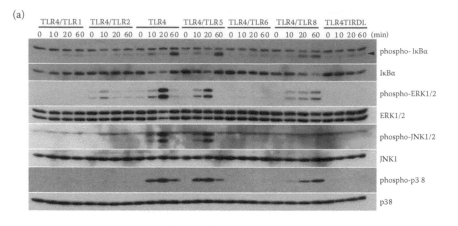

(b)

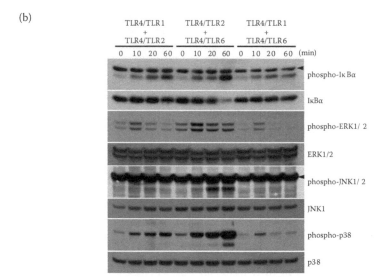

FIGURE 7.5   Analysis of the Activation of Signaling Molecules in Ligand-Regulated TLR Chimera System

Note: Ligand-induced activation of signaling molecules derived from the cytoplasmic domain of individual TLR (A) and combinations of two distinct TLRs (B). TLR4-deficient bone marrow–derived macrophages expressing an individual TLR chimera (A) or two distinct TLR chimeras (B) were stimulated with 10 ng/ml LPS for the indicated periods. Cell lysates were prepared, and the activity of NF-κB, ERK1/2, JNK1/2, and p38 was determined by immunoblotting using antibodies to unmodified and phosphorylated ERK, JNK, and p38. Note that IκBα becomes phosphorylated and degraded to activate NF-κB, and that IκBα synthesis is rapidly increased by NF-κB. TLR4TIRDL: truncated TLR4 in which the TIR domain was deleted.

Source: Modified from Nishiya, T., and DeFranco, A. L. Ligand-regulated chimeric receptor approach reveals distinctive subcellular localization and signaling properties of the Toll-like receptors. *J. Biol. Chem.* 279, 19008, 2004. With permission.

24-well plates for assays of cytokine production by ELISA, and 6-well plates for analysis by immunoblotting or reverse transcriptase PCR (RT-PCR).

4. The next day, stimulate the cells with 1–100 ng/ml LPS (usually, 10 ng/ml is enough to observe most responses).

5. After stimulation, collect supernatants for ELISA, prepare whole cell lysates for immunoblotting, or prepare total RNA for RT-PCR.

This experiment is suitable for cell surface TLRs, but not intracellular TLRs. We could not detect any responses to LPS in macrophages expressing intracellular TLR chimeras. LPS does not seem to reach intracellular TLRs.

REFERENCES

1. Akira, S., Uematsu, S., and Takeuchi, O. Pathogen recognition and innate immunity. Cell 124, 783, 2006.
2. Choe, J., Kelker, M. S., and Wilson, I. A. Crystal structure of human toll-like receptor 3 (TLR3) ectodomain. Science 309, 581, 2005.
3. O'Neill, L. A., and Bowie, A. G. The family of five: TIR-domain-containing adaptors in Toll-like receptor signaling. Nat. Rev. Immunol. 7, 353, 2007.
4. Wagner, H. The immunobiology of the TLR9 subfamily. Trends Immunol. 25, 381, 2004.
5. Barton, G. M., Kagan, J. C., and Medzhitov, R. Intracellular localization of Toll-like receptor 9 prevents recognition of self DNA but facilitates access to viral DNA. Nat. Immunol. 7, 49, 2006.
6. Marshak-Rothstein, A. Toll-like receptors in systemic autoimmune disease. Nat. Rev. Immunol. 6, 823, 2006.
7. Zhang, H., Tay, P. N., Cao, W., Li, W., and Lu, J. Integrin-nucleated Toll-like receptor (TLR) dimerization reveals subcellular targeting of TLRs and distinct mechanisms of TLR4 activation and signaling. FEBS Lett. 532, 171, 2002.
8. Takeuchi, O., Kawai, T., Muhlradt, P. F., Morr, M., Radolf, J. D., Zychlinsky, A., Takeda, K., and Akira, S. Discrimination of bacterial lipoproteins by Toll-like receptor 6. Int. Immunol. 13, 933, 2001.
9. Ozinsky, A., Underhill, D. M., Fontenot, J. D., Hajjar, A. M., Smith, K. D., Wilson, C. B., Schroeder, L., and Aderem, A. The repertoire for pattern recognition of pathogens by the innate immune system is defined by cooperation between toll-like receptors. Proc. Natl. Acad. Sci. USA 97, 13766, 2000.
10. Medzhitov, R., Preston-Hurlburt, P., and Janeway, C. A., Jr. A human homologue of the Drosophila Toll protein signals activation of adaptive immunity. Nature 388, 394, 1997.
11. Poltorak, A., Smirnova, I., Clisch, R., and Beutler, B. Limits of a deletion spanning Tlr4 in C57BL/10ScCr mice. J. Endotoxin Res. 6, 51, 2000.
12. Qureshi, S. T., Lariviere, L., Leveque, G., Clermont, S., Moore, K. J., Gros, P., and Malo, D. Endotoxin-tolerant mice have mutations in Toll-like receptor 4 (Tlr4). J. Exp. Med. 189, 615, 1999.
13. Poltorak, A., Smirnova, I., He, X., Liu, M. Y., Van Huffel, C., McNally, O., Birdwell, D., Alejos, E., Silva, M., Du, X., Thompson, P., Chan, E. K., Ledesma, J., Roe, B., Clifton, S., Vogel, S. N., and Beutler, B. Genetic and physical mapping of the LPS locus: identification of the toll-4 receptor as a candidate gene in the critical region. Blood Cells Mol. Dis. 24, 340, 1998.

14. Poltorak, A., He, X., Smirnova, I., Liu, M. Y., Van Huffel, C., Du, X., Birdwell, D., Alejos, E., Silva, M., Galanos, C., Freudenberg, M., Ricciardi-Castagnoli, P., Layton, B., and Beutler, B. Defective LPS signaling in C3H/HeJ and C57BL/10ScCr mice: mutations in Tlr4 gene. Science 282, 2085, 1998.

15. Nishiya, T., and DeFranco, A. L. Ligand-regulated chimeric receptor approach reveals distinctive subcellular localization and signaling properties of the Toll-like receptors. J. Biol. Chem. 279, 19008, 2004.

16. Nishiya, T., Kajita, E., Miwa, S., and DeFranco, A. L. TLR3 and TLR7 are targeted to the same intracellular compartments by distinct regulatory elements. J. Biol. Chem. 280, 37107, 2005.

17. Latz, E., Schoenemeyer, A., Visintin, A., Fitzgerald, K. A., Monks, B. G., Knetter, C. F., Lien, E., Nilsen, N. J., Espevik, T., and Golenbock, D. T. TLR9 signals after translocating from the ER to CpG DNA in the lysosome. Nat. Immunol. 5, 190, 2004.

18. Schneider, D. S., Hudson, K. L., Lin, T. Y., and Anderson, K. V. Dominant and recessive mutations define functional domains of Toll, a transmembrane protein required for dorsal-ventral polarity in the Drosophila embryo. Genes Dev. 5, 797, 1991.

19. Takeshita, S., Kaji, K., and Kudo, A. Identification and characterization of the new osteoclast progenitor with macrophage phenotypes being able to differentiate into mature osteoclasts. J. Bone Miner. Res. 15, 1477, 2000.

20. Yee, J. K., Friedmann, T., and Burns, J. C. Generation of high-titer pseudotyped retroviral vectors with very broad host range. Methods Cell Biol. 43 (Pt. A), 99, 1994.

21. Onishi, M., Kinoshita, S., Morikawa, Y., Shibuya, A., Phillips, J., Lanier, L. L., Gorman, D. M., Nolan, G. P., Miyajima, A., and Kitamura, T. Applications of retrovirus-mediated expression cloning, Exp. Hematol. 24, 324, 1996.

22. Akashi, S., Shimazu, R., Ogata, H., Nagai, Y., Takeda, K., Kimoto, M., and Miyake, K. Cutting edge: cell surface expression and lipopolysaccharide signaling via the toll-like receptor 4-MD-2 complex on mouse peritoneal macrophages. J. Immunol. 164, 3471, 2000.

23. Nagai, Y., Akashi, S., Nagafuku, M., Ogata, M., Iwakura, Y., Akira, S., Kitamura, T., Kosugi, A., Kimoto, M., and Miyake, K. Essential role of MD-2 in LPS responsiveness and TLR4 distribution. Nat. Immunol. 3, 667, 2002.

8 Engagement of Toll-Like Receptors Modulates Chemokine Receptor Signaling

Chong-Shan Shi and John H. Kehrl

CONTENTS

8.1 INTRODUCTION

Toll-like receptor (TLR) family members are pattern recognition receptors that function to initiate innate immune responses against many microbial pathogens, and consequently trigger the induction of adaptive immune responses.[1] TLRs are receptors prominently expressed on antigen-presenting cells, such as macrophages and dendritic cells, cells critical for initiating immune responses. TLRs are also found on T and B lymphocytes, and their respective ligands affect T and B cell function by modulating cellular proliferation, differentiation, and cytokine secretion.[2–4] TLR signaling triggers activation of the nuclear factor (NF)-κB pathway, stimulates mitogen-activated protein kinases (MAPKs) leading to AP1 activation, and induces type I interferon production.[1]

    Chemokine and chemoattractant receptors play a critical role in the migration of developing immune cells into lymphoid organs and inflammatory sites, and help to organize secondary immune tissue anatomic structures. Chemokine and chemoattractant receptors use heterotrimeric G proteins as secondary messengers. Heterotrimeric G proteins consist of three subunits: α, β, and γ. There are 23 α, 5 β, and 10 γ subunits. The different G_α subunits, along with their associated $G\beta\gamma$ subunits, have been grouped into the subfamilies G_i, G_q, G_s, and G_{12}. Members of the G_i subfamily and perhaps the G_q and G_{12} subfamilies mediate chemokine receptor signaling. Upon ligand binding, a G protein–coupled receptor (GPCR) such as a chemokine receptor functions as a guanine nucleotide exchange factor for the G_α subunit, causing the release of GDP and the binding of GTP in its place. The GTP-bound form of G_α reorganizes the heterotrimeric complex such that both G_α-GTP and $G_{\beta\gamma}$ can independently bind and activate primary downstream effectors. Gα subunits have an intrinsic GTPase activity that hydrolyzes GTP to GDP triggering reassembly with $G_{\beta\gamma}$ to form the inactive heterotrimer. There is a family of proteins called "regulators of G protein signaling" (RGSs) proteins, which accelerate the intrinsic rate of the G_α GTPase reaction and, as a result, downregulate GPCR signaling.[5]

The signaling pathways triggered by the engagement of TLRs and GPCRs significantly overlap as both activate NF-κB, MAPKs, and protein kinase Akt. There is also some recent evidence that demonstrates cross talk between TLR and GPCR signaling. For example, G_s-induced increases in adenosine 3′,5′-cyclic monophosphate (cAMP) modulates the cytokine expression pattern induced in Raw 264.7 macrophages by LPS.[6] Macrophages isolated from CXCR4-deficient mice exhibit reduced MyD88-dependent NF-κB activation, whereas p38 MAPK and JNK activation are increased.[7] Another study showed that stimulation of dendritic cells with either TLR3 or TRL4 ligands altered the expression of chemokine receptors, G_α subunits, and RGS proteins in dendritic cells, thereby altering the response of dendritic cells to chemokine stimulation.[8] Similarly, stimulation of B cells through TLR4 enhances the expression of chemokine receptors and homing receptors, reduces *Rgs1*, and enhances *Gnai2* expression (J. Kehrl, unpublished data). Because other chapters focus on TLR signaling, the major emphasis in the following sections is on GPCR signaling in lymphocytes, where the most prominent GPCRs are chemokine receptors. The last section details some methods useful for examining TLR signaling in lymphocytes and other immune cells.

8.2   CHEMOKINE RECEPTOR SIGNALING

Lymphocytes and other immune cells express a broad array of chemokine receptors, and exposure of lymphocytes to the appropriate chemokine elicits a number of measurable responses, including a transient increase in intracellular calcium levels; the activation of small GTPases, such as Rho and Rac; and transient increases in the phosphorylation of numerous proteins, including extracellular receptor kinases (ERKs), AKT, Src family kinases, and Pyk2. Chemokines also trigger cell migration.[9,10]

8.2.1   CELL PREPARATION

Specific lymphocyte subsets from mice (or humans) can be isolated as follows for Western blotting, for studies of chemokine signaling, and for cellular migration. The focus is on the preparation of lymphocytes from mice. Primary and secondary lymphoid tissues in mice can be identified with the help of a basic immunology text-book such as *Fundamental Immunology, Roitt's Essential Immunology*, or *Current Protocols in Immunology*. The following is a brief protocol.[9,11]

Protocol 8.1: Isolation of Lymphocytes

1. Sacrifice mice, and harvest lymphoid organs such as the thymus, peripheral lymph nodes, mesenteric lymph nodes, Peyer's patches, or spleen.

2. Disrupt the lymphoid organs of interest by gentle teasing and disruption with forceps, needles, or scissors in phosphate-buffered saline (PBS)/1% bovine serum albumin (BSA), and then filter through 70 mm nylon mesh (BD Falcon, BD Biosciences, San Jose, CA) to remove the connective tissue.

3. Peritoneal cavity cells can also be harvested by injecting 7–10 ml of cold PBS into the peritoneal cavity using a 27 gauge needle and removing the cell wash with a 19 gauge needle.

4. Bone marrow cells can be harvested by dissecting the femur and tibia, removing excess tissue, cutting the ends off the bones, and then flushing the marrow out with PBS containing 1% BSA through both ends of the bones.

5. Bone marrow and spleen should be depleted of red blood cells by treatment with ACK buffer (0.15 M NH_4CL, 1 M $KHCO_3$, and 0.1 M Na_2EDTA [pH 7.3]) and subsequent washing in 1× PBS.

6. The cell suspensions should be centrifuged at 1,000 rpm for 5 minutes at 4 °C, the supernatants removed, and the pellet suspended in PBS/1% BSA. A count of the suspensions should be made with a hemocytometer to determine cell numbers per milliliter.

7. Specific lymphocytes can be isolated using a fluorescent-activated cell sorter (FACS). Pellet cells isolated from tissues, and suspend in FACS buffer (PBS [pH 7.4]/1% BSA fraction V). Stain cells for specific cell markers that identify a subset of cells, such as follicular cells ($B220^+/CD21^+/CD23^+$), marginal zone B cells ($B220^+/CD21^+/CD23^-$), or T cells ($CD4^+$ or $CD8^+$). After a brief incubation at 4 °C, wash cells three times with FACS buffer and suspend at 1×10^7/ml in FACS buffer. The cells can then be sorted by flow cytometry. Obtaining significant numbers of cells with this method can be problematic. However, if you are interested in a population that represents a small percentage of cells, it may be the best way to obtain significant numbers of the population to use. After sorting, the cells should be resuspended in RPMI 1640/10% fetal calf serum (FCS) and incubated at 37 °C for 1 hour prior to any signaling or migration assay.

8. Alternatively, cell populations can be obtained by negative selection by magnetic bead separation. Purify cells from bulk populations by staining cells with specific monoclonal antibodies and negatively selecting those cells, leaving an enriched cell population. For example, to obtain B cells, suspend the cell population in FACS buffer (PBS [pH 7.4]/1% BSA fraction V) and stain with biotinylated Mab CD4, CD8, GR-1, Mac-1, Ter119, CD11c,

and DX5 to remove T cells, macrophages, erythrocytes, natural killer cells, granulocytes, and dendritic cells. Incubate cells at 4 °C for 15 minutes, wash with PBS, and suspend in FACS buffer. Add Dynabeads M-280 strepta- vidin beads (Dynal, Carlsbad, CA) washed twice in PBS and twice with FACS buffer to the cell suspensions, and incubate at 4 °C, rotating slowly for 15 minutes. The ratio of beads to cells is determined according to the manufacturer's protocol. Attach the suspensions to the magnet (Dynal), and allow them to separate. Collect the nonadherent suspension, reapply to the magnetic source, and collect the nonadherent cells in suspension once again. Wash, count, suspend in RPMI 1640 containing 10% FCS, and incubate this cell population at 37 °C for 1 hour prior to any assay.

8.2.2   CA^{2+} INFLUX

Measurement of Ca^{2+} influx has been used to examine changes in intracellular Ca^{2+} in lymphocytes. The following method can be used for chemokine-triggered intra- cellular Ca^{2+} influx. The effects of pretreatment with TLR ligands on chemokine- induced increases in intracellular Ca^{2+} can be assessed.[11]

Protocol 8.2: Measuring Calcium Influx in Lymphocytes

1. Wash cells in Hank's buffered saline solution (HBSS) with calcium (1.26 mM) and magnesium (1 mM) (Biosource International, Camarillo, CA), 10 mM HEPES, and 1% FCS, and then suspend at 1×10^7 cells/ml.
2. Add the fluorescent calcium probe Indo-1 (indo-1/acetoxymethyl ester) (Sigma-Aldrich, St. Louis, MO; or Molecular Probes, Eugene, OR) at a final concentration of 2 µg/ml plus pluronic detergent (Molecular probes) at a final concentration of 300 µg/ml.
3. Incubate the cells for 30 minutes at 30 °C while protected from light.
4. Wash cells with HBSS buffer and suspend at 1×10^7/ml in HBSS buffer. Cells can be stained with mAbs to cell surface markers for 10 minutes and washed three times in HBSS buffer.
5. Warm the cells at 37 °C for 3 minutes prior to stimulation.
6. To stimulate, load cells into the Time Zero module (Cyteck, Fremont, CA) and run at 1,000 cells per second. Collect baseline for 30 seconds, inject a sham of 50 µl of HBSS, and, at 60 seconds, inject the stimulant, such as a chemokine.
7. Perform the measurements on a flow cytometer (i.e., FACSVantage, BD, Franklin Lakes, NJ) equipped with an argon laser tuned to 488 nm and a krypton laser tuned to 360 nm. Analyze Indo-1 fluorescence at emission set- tings 390 and 530, with a slit width of 20, for bound and free probes, respec- tively. Data can be analyzed using FlowJo software (Tree Star, Ashland, OR). Results are best shown as ratio fluorescence (violet/blue).

8.2.3   Rho, Rac, and Cdc42 Activity

The activation of members of the Rho small GTPase family such as Rho, Rac, Rap1, and Cdc42 is critical for cell migration and also for function in TLR signaling.[12] Now, nonradioactive methods are available to detect the activity of RhoGTPases in lymphocytes.[13] The principle of this kind of assay is to use a specific recombinant protein to pull down GTP-bound form RhoGTPases from cellular lysates, and then perform an immunoblot to detect the interacted RhoGTPases that are GTP-bound form.

Protocol 8.3: Determination of Rho Protein Activity in Lymphocytes

1. Resuspend cells at 1×10^7/ml.
2. Add the chemokine (i.e., CXCL12 [100 ng/ml], CXCL13 [1,000 ng/ml], or CCL19 [250 ng/ml] [R&D Research, Minneapolis, MN]) to the culture at a specific time point. Harvest, centrifuge, and use the cells for making lysates by adding 1 ml precold buffer (50 mM Tris [pH 7.2], 1% Triton X-100, 0.5% sodium deoxycholate, 0.1% SDS, 500 mM NaCl, 10 mM MgCl$_2$, 10 µg/ml each of leupeptin and aprotinin, and 1mM PMSF), and vortexing and spinning lysate at 14,000 rpm 2–4 minutes at 4 °C.
3. Collect the cleared lysate, save about 10% of the lysate to run an immunoblot, and to the rest of the lysate immediately add Rhotekin RBD agarose beads (Upstate Technology, Lake Placid, NY) to detect Rho GTP-bound form or PAK-1 PBD agarose beads (Upstate Technology) to detect GTP-bound Rac or Cdc42 (15–20 µl/each sample).
4. Slowly rotate the tube with lysate and beads at 4 °C for 60–90 minutes.
5. Wash the beads with wash buffer (50 mM Tris [pH 7.2], 1% Triton X-100, 150 mM NaCl, 10 mM MgCl$_2$, 10 µg/ml each of leupeptin and aprotinin, and 1mM PMSF) 0.5 ml four times.
6. Remove the wash buffer, add SDS sample buffer (with DTT), and heat at 95 °C for 10 minutes. Run 13% SDS-PAGE.
7. Add SDS sample buffer to total cellular lysate saved from step 3, and heat at 95 °C for 10 minutes. Run 13% SDS-PAGE.
8. Transfer to a membrane, and immunoblot with Rho, Rac, or Cdc42 antibody (Upstate Technology) to detect protein levels.
9. The Rho, Rac, or Cdc42 eluted from beads is GTP bound, while the Rho, Rac, or Cdc42 present in total cell lysate contains both GTP- and GDP-bound protein. The amount of GTP-bound small GTPase should be normalized to the amount of small GTPase present in the total cell lysate.

8.2.4   ERK, Akt, and Pyk2 Phosphorylation

The phosphorylation of ERK and Akt occurs within minutes of exposure of lymphocytes to chemokines.[14] Pyk2 is a nonreceptor protein tyrosine kinase, which also is phosphorylated following chemokine stimulation.[15] TLR signaling can also lead to the phosphorylation of ERK, Akt, and Pyk2.[1,16,17] Therefore, checking the phosphorylation levels of these proteins is a useful way to observe cross talk between TLRs and chemokine receptor signaling.

Protocol 8.4: Determination of ERK, Akt, and Pyk2 Phosphorylation in Lymphocytes

1. Isolate the lymphocytes as described previously and make a single cell suspension of 1×10^7 cells/ml in PBS/1% BSA media. Cells should be incubated for 60 minutes at 37 °C prior to stimulation.
2. Add the chemokine (i.e., CXCL12 [100 ng/ml], CXCL13 [1,000 ng/ml], or CCL19 [250 ng/ml] [R&D Research, Minneapolis, MN]) to the culture. At different time points, harvest, centrifuge, and make cell lysates by adding buffer (20 mM HEPES [pH 7.4], 2 mM EGTA, 50 mM β-glycerophosphate, 1% Triton X-100, and 10% glycerol, plus protease inhibitors and 2 mM sodium pervanadate) and vortexing. Maintain them on ice for 20 minutes.
3. Centrifuge lysates for 10 minutes at 14,000 rpm at 4 °C, collect the lysate supernatant, and determine the protein concentration (Bio-Rad protein assay; Bio-Rad, Hercules, CA). Lysates can be used directly or kept at −80 °C for later use.
4. Fractionate the samples (25–100 μg protein) by SDS-PAGE and immunoblot. For example, transfer to pure nitrocellulose membranes, block the membranes with 10% milk in Tris-buffered saline (TBS), and incubate with the appropriate dilution of antiphosphospecific antibody (Cell Signaling Technology, Danvers, MA) in 5% nonfat dry milk in TBS with 0.1% Tween 20 at 4 °C overnight with gentle shaking. By carefully choosing antibodies, pPyk2, pAKT, and pERK can be examined simultaneously. Wash the blot three times with TBS plus 0.1% Tween 20 for 20 minutes. Use either a directly HRP-conjugated secondary antibody for 1 hour at room temperature or a biotinylated secondary with a third strepavidin–HRP-conjugated antibody to detect the primary antibody. Between steps, wash the blots as described earlier.
5. Detect proteins by a standard chemiluminescence assay.
6. Subsequently strip the blot and reprobe with Pyk2-, Akt-, and ERK-specific antibodies to ensure equal loading of the lanes.

8.2.5 INOSITOL PHOSPHATES

Chemokine receptors increase intracellular inositol phosphate levels, although not as efficiently as do GPCRs, which signal using members of the Gq subfamily of heterotrimeric G protein.[18] The following protocol has been used with B cells, 293 cells, and COS cells.

Protocol 8.5: Determination of Inositol Phosphates in Lymphocytes

1. Prepare lymphocytes or a lymphocyte cell line (10^6/ml), and wash twice with inositol-free RPMI 1640 medium.
2. Cells are labeled with myo-[^{3}H]inositol (4 µCi/ml; Amersham Biosciences, Piscataway, NJ) in inositol-free RPMI 1640 medium supplemented with 10% fetal calf serum for 14–16 hours. Option: use inositol-free RPMI 1640 medium supplemented with 2% or 5% fetal calf serum to reduce the basal production of inositol phosphates.
3. Wash the cells twice with the same medium without serum.
4. Add 1 ml of the medium with 100 µl of 200 mM LiCl (freshly made and filtered) at 20 mM final concentration.
5. Incubate the cell for 20–30 minutes.
6. Stimulate the cells with chemokines at appropriate concentrations for various durations.
7. Harvest the cells by quickly spinning at 4 °C. Resuspend the cells with 0.5 ml of 20 mM formic acid to terminate signaling and to extract the inositol phosphates in the cells.
8. Place the tube with cells on ice for 30 minutes, then centrifuge the tube at 14,000 rpm for 5 minutes at 4 °C. Collect the supernatant. Option: repeat steps 7 and 8 once.
9. The supernatant is saved on the ice for the next processes. The pellet can be used to detect protein expression by SDS-PAGE.
10. Neutralize the supernatant to pH 7.5 with about 150 µl of a solution (7.5 mM HEPES and 150 mM KOH), if the total supernatant volume is about 1 ml. (Notice: pH 7.5 is very critical for the next steps.)
11. Centrifuge the supernatant at 14,000 rpm for 5 minutes at room temperature to clear the nonsoluble portion.
12. Equilibrate a 0.5 ml Dowex AG1-X8 column (100–200 mesh format, from Catalog no. 140-1444, Bio-Rad, Hercules, CA) with 2 ml 1M NaOH, then 2 ml 1M formic acid, and then wash the column five times, each time with 5 ml water.
13. Load the neutralized supernatant from step 11 to the column drop by drop.
14. Wash the column with 5 ml water and 5 ml of wash solution (5 mM borax and 60 mM sodium formate).
15. Elute with 3 ml eluting solution (0.9 M ammonium formate and 0.1 M formic acid).
16. Collect the eluted solution (3 ml); take about 0.2 ml mixed with CytoScint 10 ml, and count in a scintillation counter.

8.2.6 CHEMOTAXIS

The most common method for measuring lymphocyte chemotaxis uses a chamber based on Boyden's original design,[19] but it has been adapted so that multiple samples can be assayed at one time. This method provides a rapid and easily quantifiable means for assessing the response of lymphocytes to a chemokine. By employing flow cytometry, the chemotaxic responses of subpopulations of cells within an input population can be measured. In addition, if an assay for an individual cell characteristic exists, this method can be adapted to examine the migratory ability of functional subsets of cells.[9]

Protocol 8.6: Lymphocyte Chemotaxis Assay

For the assay, use a 5 μm pore size chamber for both primary lymphocytes and most lymphocyte cell lines (Costar Transwell Chambers, no. 3421, Corning, Corning, NY). In the bottom of the chamber, add 600 μl of media either with or without chemoattractant. For Jurkat cells (a human T cell line), 100 ng/ml of CXCL12 is sufficient for a maximal response.

1. Place the insert into the chamber, add 100 μl of cells prewarmed to 37 °C to the inserts, and incubate cultures for 2–3 hours at 37 °C.
2. After the incubation, remove the insert, collect cells from the lower chamber, spin them down, and suspend them in 300 μl of FACS buffer.
3. On a flow cytometer, collect data for 1 minute for each sample at the high flow rate (important for the calculations).
4. In order to calculate the percentage migration, a 100 μl aliquot from the input cell sample should be run on the flow cytometer like the other samples.
5. The percentage migration can be calculated by the following formula: the cells in the bottom of the chemoattractant-containing well minus the cells in the bottom without chemoattractant divided by the cells in the input sample.

8.2.7 Lymphocyte Trafficking

A number of approaches have been used to study lymphocyte trafficking in vivo, including imaging of explanted lymph nodes or intravital imaging using two-photon or confocal microscopes.[20,21] One approach, which does not require expensive imaging equipment but is limited to animal studies, has been to ex vivo label mouse lymphocytes and then reintroduce them into a donor animal for tracking using either the flow cytometric analysis of cells derived from lymphoid tissues or immunohistochemistry. This approach is most useful for examining the in vivo trafficking of genetically modified lymphocytes or lymphocytes treated with pharmaceutical agents, whose effects are not readily reversible.[9,11]

Protocol 8.7: Lymphocyte Trafficking Assay

1. To label high numbers of cells with CFSE (Sigma-Aldrich, St. Louis, MO; or Molecular Probes, Eugene, OR), resuspend the cells at a concentration of 50×10^6/ml in HBSS (without serum). For lower cell numbers, resuspend the lymphocytes in PBS containing 5% FCS at 0.5 to 10×10^6 cells/ml. FCS is

needed at the low cell concentration to buffer the toxic effect of the concentrated dye.

2. Dilute the stock CFSE dye to 50 µM in PBS, add 110 µl of dye solution (final concentration is 1–5 µM) per ml of cells, and mix rapidly by pipetting.

3. Incubate for 5 minutes at room temperature, and wash with 10 volumes of PBS 5% FCS. Centrifuge cells for 5 minutes at 300g at 20 °C, and remove supernatant. Wash three additional times or until wash buffer is clear.

4. To label lymphocytes with lipophilic dyes, resuspend the cells at $5–10 \times 10^6$ cells/ml in buffered salt solution (HBSS).

5. Add the dye to the cell solution so that its final concentration is 0.5–50 µM. Incubate the cells at room temperature for 5 minutes.

6. Remove excess dye by centrifuging the cells for 5 minutes at 300g at 20 °C. Remove the supernatant and add 10 volumes of HBSS with 5% FBS. Repeat the wash three times.

7. To transfer the labeled lymphocytes, suspend them in PBS and inject $10–40 \times 10^6$ cells in 0.2 ml intravenously into the lateral tail vein of a mouse.

8. Lymphoid organ cells of recipient mice are typically harvested 30–120 minutes after transfer, counted, and subjected to flow cytometric analysis. Because CFSE acquires identical spectral characteristics as fluorescein when carboxyl groups are cleaved by cytoplasmic esterases, it is excited efficiently by the 488 nm argon laser standard in most flow cytometers. This allows the concurrent use of other fluorochromes such as phycoerythrin (PE) or allophycocyanin (APC). Block single-cell suspensions of 1×10^6 cells per sample with the appropriate blocking reagents (such as 5% BSA/PBS or CD16/CD32 Fc block; if a secondary monoclonal is needed to detect the primary monoclonal antibody, the cells should be blocked with 5% serum of the source of the secondary antibody in a protein-blocking solution) on ice for 15 minutes, wash, and then incubate with PE- or APC-labeled mAb in FACS buffer. After 15 minutes of incubation on ice, wash the samples three times with FACS buffer and perform analysis on a flow cytometer. If the analysis is to occur more than 24 hours after transfer, a higher concentration of CFSE should be used (5 µM) in the initial labeling. Despite their loss of fluorescence with time, CFSE-labeled cells can be detected for several months following transfer. The accumulation of wild-type cells in various lymphoid organs can be compared to that of genetically modified cells.

9. The positioning of the transferred cells in lymphoid tissues can be determined by fluorescence microscopy of tissue sections. Use harvested lymphoid tissues to make frozen tissue blocks in Tissue Tek OCT compound (Fisher Scientific, Waltham, MA). Fix tissue sections in 4% paraformaldehyde in 0.1 M phosphate buffer (pH 7.4), wash twice in PBS, and then incubate in 20% sucrose PBS at 4 °C for 1 hour. After rinsing in cold PBS, snap-freeze the tissues in Tissue Tek OCT compound (Fisher). Tissue sections (8 µm) can be cut from the snap-frozen block using a cryostat. Rehydrate thawed sections in PBS, fix again in 4% paraformaldehyde as described earlier, wash in PBS, and then quench with 50 mM NH$_4$Cl in PBS (quenching of debatable value). The tissue sections can then be stained with the APC- or PE-labeled anti-

bodies of interest. Note that no organic fixatives can be used (e.g., alcohol, acetone, or xylene), as they will remove the dye from the tissue. Only water-based fixatives can be used such as paraformaldehyde. Additionally, paraffin embedding cannot be used.

8.3 TLR SIGNALING

Because TLR signaling is covered in greater depth in other chapters, only a brief review of the preparation of immune cells for TLR signaling and a basic outline of the assays commonly used to detect TLR signaling are included here. The interactions between GPCR signaling and Toll-like receptor signaling can be monitored by simultaneously or sequentially applying the GPCR ligands and TLR ligands to immune cells.

8.3.1 ISOLATION OF MACROPHAGES AND DENDRITIC CELLS

Lymphocytes, dendritic cells, and macrophages can be prepared from mice. Negative selection is best for isolating mouse B and T cells from mouse lymphoid organs. The methods have been described above. LPS and CpG DNA are suitable ligands for triggering TLR4 and TLR9 signaling in B cells. Dendritic cells (DCs) and macrophages express multiple TLRs. RAW 264.7 cells, a mouse macrophage cell line, are useful for checking Toll-like receptor signaling triggered by TLR3 or TLR4. In addition, RAW 264.7 cells are readily transfected with FuGENE-HD Transfection Reagent (Roche Applied Science, Basel, Switzerland), following the protocol from the manufacturer. Bone marrow–derived dendritic cells and macrophages can be prepared from mice as follows.[8,22]

Protocol 8.8: Isolation of Macrophages and Dendritic Cells

1. Mice, BALB/C, approximately 7 weeks old, are used for bone marrow culture. Remove all muscle tissues from the femora and tibias with gauze. Place the bones in a 60 mm dish with 70% alcohol for 1 minute, wash twice with PBS, and transfer the bones into a fresh dish with RPMI 1640.
2. Cut both ends of the bones with scissors in a dish, and then flush out the marrow using 2 ml of RPMI 1640 with a syringe with a 25 gauge needle.
3. Resuspend the marrow tissue, and pass through a nylon mesh to remove small pieces of bone and debris. Next, remove residual red blood cells with ACK lysis buffer.
4. The bone marrow cells (2×10^5 cells/ml) are cultured in 100 mm^2 culture dishes (10 ml/dish) in complete medium containing 2 ng/ml GM-CSF. At day 3, add another 10 ml of fresh complete medium containing 2 ng/ml GM-CSF. On day 6 of the culture, change half of the medium. On day 8 of the culture, harvest the nonadherent DCs and loosely adherent DCs by gentle pipetting.

5. Transfer the harvested DCs to a new dish, and culture 2–4 hours in fresh medium without GM-CSF. Then stimulate the cells with TLR ligands for suitable time periods using a range of concentrations.
6. Harvest TLR ligand treated cells, and check TLR signaling (see below).
7. Bone marrow–derived macrophages are prepared using a similar method with slight modifications. Transfer the collected bone marrow cells to a 100 mm Petri dish with fresh complete medium containing 10ng/ml M-CSF. Change half of the medium on day 4. Harvest the macrophages on day 7 for stimulation and detection of TLR signaling.[23]

8.3.2   TLR SIGNALING ASSAYS

For screening purposes, a set of TLR ligands is available (Apotech, Epalinges, Switzerland). Pam_3CSK_4 is a synthetic tri-palmitoylated lipopeptide, an analog of the N-terminal portion of bacterial lipoprotein, which potently activates macrophages by triggering a heterodimer of TLR1 and TLR2 (10 to 100 ng/ml). Poly (I:C) is a synthetic mimetic of viral double-stranded RNA, which targets TLR3 (25 to 100 μg/ml). Lipopolysaccharide (LPS) is isolated and purified from *E. coli*. LPS is used for activating TLR4 (0.02 to 1μg/ml). Flagellin is generally isolated from *Salmonella typhimurium*. Flagellin activates TLR5 (10 to 100 ng/ml). Macrophage stimulatory lipopeptide 2 (MALP-2) is isolated from *Mycoplasma fermentans* and is used to activate TLR6 (10 to 100 ng/ml). Poly (U) is a simple motif single-stranded RNA that can activate TLR7 or TLR8 (0.1 to 10 μg/ml). Synthetic oligodeoxynucleotides (ODN) containing an unmethylated deoxycytosine-deoxyguanosine (CpG) motif mimic the effects of bacterial DNA and activate TLR9 (0.1 to 10 μg/ml).

Ligand bond TLRs trigger the phosphorylation of several proteins, which can serve as indicators of pathway activation. Specific antibodies against phosphorylated IRAK, TAK1, IKKα/β, IκB-α, and MAPKs are available (Cell Signaling Technology, Danvers, MA), which can be used for immunoblotting to detect these specific protein phosphorylations. TLR signaling also activates the NF-κB pathway by triggering the degradation of IκB-α, which occurs 10–20 minutes after ligand exposure. Immunoblotting with an antibody that recognizes IκB-α can be used to determine the expression levels of IκB-α. Transcriptional reporter assays are also useful for monitoring the NF-κB and interferon pathways. Commonly, luciferase reporter genes are used. Cells should also be co-transfected with internal control constructs such as *Renilla* luciferase (phRL-TK, Promega, Madison, WI). The cells can be seeded in 48-well plates. The reporter and control constructs (20–100 ng/each well) are transfected for 24 hours using standard transfection techniques. TLR ligands are used to stimulate the cells for at least 5 hours. A useful assay kit (Dual-Glo™ luciferase assay kit, Promega), allows reporter gene and control gene activities to be measured using a luminometer. Stimulation of DCs and macrophages with various TLR ligands leads to the production of inflammatory cytokines, such as tumor necrosis factor (TNF), interleukin (IL)-1β, IL-6, IL-10, IL-12, IL-23, and type I interferons. Companies such as BD Biosciences (San Jose, CA) and eBioscience (San Diego, CA) have useful kits for enzyme-linked immunosorbent assays (ELISA) to detect the productions of these cytokines.

REFERENCES

1. Kawai, T., and Akira, S. TLR signaling. *Cell Death Differ.* 13, 816, 2006.
2. Kabelitz, D. Expression and function of Toll-like receptors in T lymphocytes. *Curr. Opin. Immunol.* 19, 39, 2007.
3. Pasare, C., and Medzhitov, R. Control of B-cell responses by Toll-like receptors. *Nature* 438, 364, 2005.
4. Ruprecht, C. R., and Lanzavecchia, A. Toll-like receptor stimulation as a third signal required for activation of human naive B cells. *Eur. J. Immunol.* 36, 810, 2006.
5. Kehrl, J. H. Chemoattractant receptor signaling and the control of lymphocyte migration. *Immunol. Res.* 34, 211, 2006.
6. Natarajan, M., Lin, K. M., Hsueh, R. C., Sternweis, P. C., and Ranganathan, R. A global analysis of cross-talk in a mammalian cellular signalling network. *Nat. Cell Biol.* 8, 571, 2006.
7. Ness, T. L., Ewing, J. L., Hogaboam, C. M., and Kunkel, S. L. CCR4 is a key modulator of innate immune responses. *J. Immunol.* 177, 7531, 2006.
8. Shi, G. X., Harrison, K., Han, S. B., Moratz, C., and Kehrl, J. H. Toll-like receptor signaling alters the expression of regulator of G protein signaling proteins in dendritic cells: implications for G protein-coupled receptor signaling. *J. Immunol.* 172, 5175, 2004.
9. Moratz, C., and Kehrl, J. H. In vitro and in vivo assays of B-lymphocyte migration. *Methods Mol. Biol.* 271, 161, 2004.
10. Shi, C. S., and Kehrl, J. H. PYK2 links G(q)alpha and G(13)alpha signaling to NF-kappa B activation. *J. Biol. Chem.* 276, 31845, 2001.
11. Moratz, C., Harrison, K., and Kehrl, J. H. Regulation of chemokine-induced lymphocyte migration by RGS proteins. *Methods Enzymol.* 389, 15, 2004.
12. Ruse, M., and Knaus, U. G. New players in TLR-mediated innate immunity: PI3K and small Rho GTPases. *Immunol. Res.* 34, 33, 2006.
13. Tan, W., Martin, D., and Gutkind, J. S. The Galpha13-Rho signaling axis is required for SDF-1-induced migration through CXCR4. *J. Biol. Chem.* 281, 39542, 2006.
14. Tilton, B., Ho, L., Oberlin, E., Loetscher, P., Baleux, F., Clark-Lewis, I., and Thelen, M. Signal transduction by CXC chemokine receptor 4. Stromal cell-derived factor 1 stimulates prolonged protein kinase B and extracellular signal-regulated kinase 2 activation in T lymphocytes. *J. Exp. Med.* 192, 313, 2000.
15. Dikic, I., Dikic, I., and Schlessinger, J. Identification of a new Pyk2 isoform implicated in chemokine and antigen receptor signaling. *J. Biol. Chem.* 273, 14301, 1998.
16. Gelman, A. E., LaRosa, D. F., Zhang, J., Walsh, P. T., Choi, Y., Sunyer, J. O., and Turka, L. A. The adaptor molecule MyD88 activates PI-3 kinase signaling in CD4+ T cells and enables CpG oligodeoxynucleotide-mediated costimulation. *Immunity* 25, 783, 2006.
17. Hazeki, K., Masuda, N., Funami, K., Sukenobu, N., Matsumoto, M., Akira, S., Takeda, K., Seya, T., and Hazeki, O. Toll-like receptor-mediated tyrosine phosphorylation of paxillin via MyD88-dependent and -independent pathways. *Eur. J. Immunol.* 33, 740, 2003.
18. Shi, C. S., Lee, S. B., Sinnarajah, S., Dessauer, C. W., Rhee, S. G., and Kehrl, J. H. Regulator of G-protein signaling 3 (RGS3) inhibits Gbeta1gamma 2-induced inositol phosphate production, mitogen-activated protein kinase activation, and Akt activation. *J. Biol. Chem.* 276, 24293, 2001.

19. Boyden, S. The chemotactic effect of mixtures of antibody and antigen on polymorpho-nuclear leucocytes. *J. Exp. Med.* 115, 453, 1962.
20. Cahalan, M. D., Parker, I., Wei, S. H., and Miller, M. J. Real-time imaging of lympho-cytes in vivo. *Curr. Opin. Immunol.* 15, 372, 2003.
21. Delon, J., Stoll, S., and Germain, R. N. Imaging of T-cell interactions with antigen presenting cells in culture and in intact lymphoid tissue. *Immunol. Rev.* 189, 51, 2002.
22. Inaba, K., Inaba, M., Romani, N., Aya, H., Deguchi, M., Ikehara, S., Muramatsu, S., and Steinman, R. M.. Generation of large numbers of dendritic cells from mouse bone marrow cultures supplemented with granulocyte/macrophage colony-stimulating fac-tor. *J. Exp. Med.* 176, 1693, 1992.
23. Stanley, E. R. Murine bone marrow-derived macrophages. *Methods Mol. Biol.* 75, 301, 1997.

9 Role of Toll-Like Receptor Signaling in Central Nervous System Infections

Tammy Kielian

CONTENTS

9.1 INTRODUCTION

9.1.1 OVERVIEW OF CENTRAL NERVOUS SYSTEM (CNS) BACTERIAL INFECTIONS

Bacterial infections can be generally subdivided into two categories based on the anatomical location of lesions. Bacterial meningitis, as the name implies, involves infection of the subarachnoid space and can be caused by a wide array of organisms. *Neisseria meningitidis*, *Streptococcus pneumoniae*, and *Haemophilus influenzae* are among the leading etiologic agents of community-acquired meningitis in humans.[1,2] Despite advances made in vaccination and treatment strategies, bacterial meningitis remains associated with a significant mortality rate and incidence of neurological sequelae, particularly in very young and elderly patients. Long-term effects resulting from meningitis include hearing loss, hydrocephalus, and sequelae associated with parenchymal damage including memory loss, cerebral palsy, learning disabilities, and seizures.[3,4] Bacterial meningitis elicits a complex myriad of pathophysiological changes, many of which have been attributed to an excessive host antibacterial immune response. For example, besides the direct damage induced by pathogens, the host antibacterial response elicited during the acute phase of bacterial meningitis can be detrimental to neurons and other glia in the CNS due to the toxic effects of cytokines, chemokines, proteolytic enzymes, and oxidants produced locally at the site of infection.[1,2,5,6] Work by other laboratories has implicated TLR2 as an important sensor of bacterial infection in *Streptococcus pneumoniae* meningitis leading to the subsequent release of proinflammatory mediators that have been implicated in host tissue damage.[7,8] An additional report has indicated that MyD88, one of the major adaptor proteins utilized for transducing TLR activation signals, also impacts the host response to meningitis.[9]

Brain abscesses develop in response to a parenchymal infection with pyogenic bacteria, beginning as a localized area of cerebritis and evolving into a suppurative lesion surrounded by a well-vascularized fibrotic capsule.[10–12] Brain abscesses are typified by extensive edema and tissue necrosis, and tend to localize at white-gray matter junctions where microcirculatory flow is poor.[13,14] The most common sources of brain abscess are direct or indirect cranial infections arising from the paranasal sinuses, middle ear, and teeth. Other routes include seeding of the brain from distant sites of infection in the body (i.e., endocarditis) or penetrating trauma to the head. The leading etiologic agents of brain abscesses are the *streptococcal* strains and *S. aureus*.[11,12] Even with the recent advances made in imaging techniques, the clinical diagnosis of brain abscess is often delayed since patients may present with nonspecific clinical signs.[15] Another factor complicating diagnosis is the tendency of brain abscesses to appear similar to necrotic tumors on computer tomography (CT) and magnetic resonance imaging (MRI) scans. If not detected early, an abscess has the potential to rupture into the ventricular space, a serious complication with an 80% mortality rate. The current treatment for brain abscess includes surgical drainage to relieve pressure and long-term (6–8 weeks) systemic antimicrobial therapy.

In addition to the sequential progression from cerebritis to necrosis during brain abscess evolution, the activation of resident glial cells and influx of peripheral leukocytes demonstrate temporal patterns.[10,16,17] Specifically, microglial and astrocyte activation is evident immediately following the entry of bacteria into the CNS parenchyma and persists throughout abscess evolution.[18,19] Neutrophils are the initial leukocyte subset to infiltrate developing abscesses and are observed as early as 12 hours following bacterial exposure.[19,20] Macrophages and T cells are associated with lesions as they progress, with infiltrates generally more pronounced around days 3 and 7, respectively.[16,19] Beginning around 7 to 10 days post infection, a highly vascularized fibrotic wall forms around the necrotic milieu, effectively forming a barrier to contain the infection.[19,21,22] It is important to note that the kinetics of cellular activation, influx, and bordering functions represent general time frames, and there is likely overlap between each of these processes during brain abscess evolution.

Currently, we do not have a good understanding of the pathological mechanism(s) that dictate whether animals succumb to infection during brain abscess development. However, interpretation of available studies suggests a multifactorial etiology where both elevated cerebral edema and necrosis play a role.[13,14,20] Cerebral edema is generally divided into two types, vasogenic and cytotoxic, depending on the mechanism of edema formation.[23] Cytotoxic edema originates from the disruption of normal osmotic gradients across the cell membrane and the resultant influx of water into the cell and eventual cell swelling and lysis. In contrast, vasogenic edema occurs as a result of blood-brain barrier (BBB) compromise resulting in extracellular fluid accumulation. Both forms of cerebral edema can contribute to an increase in intracranial pressure (ICP); however, vasogenic edema is thought to represent the predominant response observed in brain abscess.[13,14] This pathogenic fluid dynamic in the CNS is likely one causative agent for the ensuing tissue necrosis observed in brain abscesses through the physical compression of cerebral vasculature and subsequent ischemia.

As indicated above, the cell types immediately responding to parenchymal infection in brain abscess differ from meningitis, and include resident microglia and astrocytes. These two glial populations play critical roles in the initial recognition of bacterial pathogens in the parenchyma leading to the production of chemokines and cytokines that serve to recruit and activate peripheral immune cells into the infectious milieu, respectively. Recent studies from our group and others have demonstrated that TLRs play a pivotal role in cytokine signaling in response to both intact bacteria as well as pathogen-associated molecular patterns (PAMPs).[5,6,24–33]

9.1.2 Expression of TLRs in the CNS

Studies examining TLR expression and function in the CNS have increased dramatically over recent years and have begun to evaluate their roles not only during infections but also in the context of noninfectious conditions of the CNS.[34–36] Within the CNS parenchyma, microglia and astrocytes have received the most intense scrutiny with regard to TLR expression and function. Microglia express the largest repertoire of TLRs, including TLR1–9, whereas astrocyte expression is more restricted and includes TLR2, TLR3, and TLR9, which may stem from the fact that although these cells are capable of mounting immune responses, this is likely not their primary function in the CNS (reviewed in Kielian, 2006).[29]

Our group has demonstrated that in response to *S. aureus*, a common etiological agent of brain abscess, microglia utilize TLR2 to recognize PGN derived from the bacterial cell wall, whereas responses to intact organisms involve not only TLR2 to a limited extent but also additional pattern recognition receptors (PRRs).[26,37] Others have also demonstrated that microglia are responsive to numerous PAMPs of biological significance during various CNS infections as well as the functional implications of TLR4 in the recognition of LPS from the outer cell wall of gram-negative bacteria.[5,6,28–33,38,39] Unlike microglia, astrocytes utilize TLR2 for recognizing both intact *S. aureus* as well as PGN,[24] suggesting that they possess a more limited array of PRRs to recognize pathogens, which is in agreement with their more selective expression of TLRs.[31,40]

Collectively, these studies have revealed to date that both microglia and astrocytes utilize TLRs to facilitate the recognition and cytokine responses to several PAMPs normally encountered during CNS bacterial infections. However, the repertoire of TLRs on microglia is more extensive, which is not unexpected given their bone marrow origin and functional similarity to monocytes and macrophages. Recent studies have indicated that TLRs facilitate recognition events not only of foreign pathogens but also of endogenous molecules released during injury.[34–36] This raises the concept of dual recognition and suggests that during CNS infections, bacterial PAMPs may synergize with endogenous "danger signal" molecules to heighten inflammatory responses.[41,42] Potential endogenous TLR agonists that may potentiate inflammation include heat shock proteins (HSP), extracellular matrix fragments, and β-amyloid.[41,43] Whether this synergy is mediated completely or partially by TLRs remains to be demonstrated; however, a two-hit model could help to explain why host immune responses in the CNS during bacterial infections have been implicated in disease pathophysiology due to their propensity to become dysregulated.

9.1.3 SUMMARY OF CHAPTER OBJECTIVES

Currently, the most definitive way to demonstrate the functional importance of TLR signaling is through the use of TLR-deficient or knockout (KO) mice. We have utilized this approach to demonstrate the role of TLR2 and the adaptor MyD88 in enabling primary microglia and astrocytes to recognize and respond to both intact *S. aureus* and its cell wall product, PGN.[24,26,37] This chapter will present specific techniques that can be used to examine various aspects of TLR usage in primary glia in response to CNS infectious insults. Detailed protocols for the isolation and purification of microglia and astrocytes, and methods to investigate the role of TLR-dependent signaling in regulating gap junction communication between glia, will be described. Of course, this is only a small subset of the available approaches that may be employed to investigate the functional roles of TLRs in response to PAMPs of clinical relevance to CNS pathology. Emerging themes integrating the function of TLRs and the interface with CNS homeostasis during infectious diseases are currently being investigated[34,41,44–46] and hold promise for unraveling important relationships regarding the functional interplay between TLRs, pathogens, and neuron homeostasis and survival.

9.2 GLIA CULTURES

Primary glia from age-matched TLR KO and wild-type (WT) mice brains can be studied to model some of the complex recognition effects that are likely occurring during bacterial infection in the CNS. The examination of TLR activities of glial cells in isolation provides a more simplified approach to investigate the effector functions of these CNS phagocytes in the absence of confounding effects by surrounding cell types that also express these TLRs (i.e., infiltrating peripheral leukocytes).

Several protocols exist for the isolation of microglia and astrocytes from the fetal,[47,48] neonatal,[49,50] and adult[51–53] rodent brain. It is important to acknowledge that various factors, including the age of the animal from which cells are procured and whether brain cells are initially maintained as mixed glial cultures prior to subset purification or immediately purified using magnetic bead or FACS analysis, can all potentially influence the downstream responses obtained. The protocol for the isolation of primary microglia and astrocytes from the neonatal mouse brain routinely utilized in our laboratory[24,49] is described.

Neonatal C57BL/6 wild-type and TLR-deficient mice (available from the Jackson Laboratory, Bar Harbor, ME) are used to prepare mixed glial cultures. All of the currently available TLR KO strains are on a C57BL/6 background; hence, the use of C57BL/6 mice is an acceptable control for comparisons. It is imperative that when attempting to compare glial responses from TLR KO and WT mice, the pups from both strains are at the same day of age when preparing mixed glial cultures. This will help to alleviate any differences that may occur due to different glial cell maturation states. We typically prepare WT cultures in the morning, followed by TLR KO in the afternoon to avoid any cross-contamination between the two cell types. The use of neonatal animals enhances glial cell recovery since the immature CNS is not fully myelinated.[54,55] Myelin traps cells during the isolation process and, inevitably, reduces cell recovery, requiring the use of additional animals. However, it is important to note that recent studies have demonstrated that the responses of neonatal versus adult microglia differ;[51,52] therefore, one should be aware of these differences and select the in vitro model that is most applicable to the disease paradigm studied.

Protocol 9.1: Isolation of Microglia from the Neonatal Mouse Brain

All instruments should be autoclaved prior to use, and a sterile beaker containing 70% EtOH placed in the tissue culture hood to clean instruments between animals. We typically also include a sterile beaker containing 1× PBS to briefly rinse instruments after immersion in 70% EtOH to prevent exposure of brain tissues to excess alcohol.

1. Euthanize 1–4-day old pups using an overdose of inhaled isoflurane (or an alternative inhalation anesthetic), and immediately embed in ice to retard postmortem degenerative changes. We do not utilize carbon dioxide to sacrifice neonatal animals since this approach takes considerable time due to the high affinity of oxygen for hemoglobin in the neonate. Briefly immerse euthanized pups in 70% EtOH to prevent bacterial contamination, and remove the heads. Make a superficial cut along the midline of the fragile skull to expose the brain. The skull at this early age is nearly translucent and is extremely fragile, so care should be taken to prevent damage to the brain. Working with one mouse at a time, remove the brain from the skull and place it on a 2 × 2 cm square of sterile gauze contained in a 100 mm Petri dish. Gently roll the brain on the gauze on all surfaces until the tissue is a creamy pale color, which indicates effective removal of meninges. The removal of the meninges is important because meningeal fibroblasts can quickly overwhelm cultures, leading to a reduction in glial cell recovery. Remove the brain stems, and transfer the cerebra to a 50 ml tube containing PBS/2% FBS on ice.
2. After all brain tissue has been collected, transfer the cerebra to a sterile 100 mm dish, and mince them using two sharp-edged forceps operated in a scissor-like fashion. This is continued until all large pieces of tissue have been reduced in size to approximately less than 1 mm^3. Next, suspend the tissue in 0.05% trypsin (Cellgro, Manassas, VA), and gently shake on an orbital

shaker at approximately 60 rpm at 37 °C for 20 minutes to facilitate tissue dissociation.

3. Transfer trypsinized tissue to a 50 ml tube, and gently resuspend it by trituration through a pipette in glial culture medium (GCM) consisting of complete Dulbecco's modified Eagle's medium (DMEM, 4.5 g/L glucose; Cellgro, Manassas, VA) containing 10% fetal bovine serum (Hyclone, Logan, UT); 200 mM L-glutamine; 100 U/ml penicillin, 0.1 mg/ml streptomycin, and 0.25 µg/ml fungizone (P/S/F mix; Cellgro); OPI medium supplement (oxalacetic acid, pyruvate, and insulin; Sigma-Aldrich, St. Louis, MO); and 0.5 ng/ml recombinant mouse granulocyte-macrophage colony-stimulating factor (GM-CSF; BD Pharmingen, San Diego, CA). Care should be taken to avoid air bubbles during trituration that will compromise cell viability. The cell suspension is filtered through a 70 µm mesh mounted in a cell strainer (Fisher Scientific, Waltham, MA) to remove debris and large aggregates, and the flow-through is collected in a new 50 ml tube.

4. Pellet cells by centrifugation at 1,000 rpm for 5 minutes at 4° C, resuspend the pellet in GCM, and seed the cells into 150 cm^2 flasks (Corning-Costar, Corning, NY). Typically we utilize the relationship of 4–5 brains per 150 cm^2 flask. These mixed glial cultures are grown under 95% air/6.5% CO_2 for approximately 7–10 days to reach confluency. At this point, astrocytes will appear as a tightly packed monolayer on which microglia sit, as either individual rounded cells or aggregates.

5. To recover microglia, shake the flasks overnight at 200 rpm on a rotary shaker at 37 °C. During this period, the astrocyte monolayer remains firmly attached to the flask, whereas microglia are dislodged from the astrocytic monolayer into the medium. Flasks can be secured tightly to the rotary shaker platform using bungee cords, and caps must be tightened to prevent the culture medium from becoming overtly basic due to the entry of room air. After the overnight shaking period, the supernatant containing microglia is collected, and fresh complete medium is added to the flasks and incubated another 5 to 7 days. The GM-CSF-supplemented GCM allows for continued expansion of microglia; however, we have recently demonstrated that the low doses of this growth factor used to induce microglial mitosis do not lead to overt alterations in the subsequent responsiveness of the microglia to a wide array of TLR agonists (with the exception of the TLR9 agonist CpG oligodeoxynucleotide [ODN]).[49] Therefore, low-dose GM-CSF can be considered a reliable method to achieve higher microglial yields without introducing dramatic activation artifacts. The microglial recovery procedure can be repeated up to three times, whereupon flasks can be used for astrocyte isolation. The density of microglial cultures for downstream experiments varies according to the type of culture dishes utilized. Typically, microglia are cultured at a seeding density of 2×10^5 or 2×10^6 for 96-well or 6-well plates, respectively.

Protocol 9.2: Isolation of Astrocytes from the Neonatal Mouse Brain

For purification of primary astrocytes, mixed glial cultures are prepared as in Protocol 9.1. Following the final shake-off to recover microglia, the adherent cells in the flasks are trypsinized and passaged at a 1:2 split in GCM as described above, but with two notable differences. First, GM-CSF is not added to the astrocyte culture medium to discourage further expansion of the remaining microglia, and, second, the medium is supplemented with 0.1 mM of the microglial cytotoxic agent L-leucine methyl ester (L-LME; Sigma-Aldrich, St. Louis, MO). Astrocytes are treated with L-LME for at least 2 weeks prior to use in our studies. In addition to these precautions, we also passage our primary astrocyte cultures a minimum of two times prior to use in experiments. The summation of these approaches is critical since microglia are notorious for hiding under confluent astrocyte monolayers. It is important that astrocyte cultures are not seeded too sparsely when passaging since cell growth will be significantly slowed.

Astrocyte contamination with microglia is an important consideration when performing studies with TLR ligands, especially since the latter is a potent innate immune effector cell and, even if present in low numbers, may be sufficient to mount a significant cytokine response that could mistakenly be interpreted as an astrocytic response. Based on this fact, additional steps may be included to improve astrocyte purity, such as negative selection paradigms using magnetic beads or FACS where CD11b$^+$ microglia are positively depleted.

The purity of astrocyte cultures is classically determined by immunocytochemical staining using antibodies for CD11b (BD Pharmingen, San Diego, CA) and GFAP (DAKO, Carpinteria, CA) to detect microglia and astrocytes, respectively. The purity of astrocyte cultures prepared in this manner is routinely > 90%. An alternative and more sensitive method to investigate astrocyte culture purity is via quantitative reverse transcription PCR (qRT-PCR) using GFAP- and CD11b-specific primers. In this approach, initial experiments using both immunocytochemistry and qRT-PCR have to be performed in parallel to correlate mRNA ratios with actual cellular ratios.

9.3 ASSESSING INFLUENCE OF TLR SIGNALING ON GLIAL GAP JUNCTION COMMUNICATION

9.3.1 OVERVIEW OF GAP JUNCTIONS

Gap junctions serve as low-resistance intercellular conduits that allow for the direct transfer of select small molecular-weight molecules (< 1 kDa) including ions, metabolites precursors, and second messengers. Gap junctions are composed of hydrophilic transmembrane pores directly linking the cytoplasm of neighboring cells.[56–58] Each cell contributes a hemichannel (termed a "connexon") composed of a hexameric assembly of proteins referred to as "connexins" (Cx). The observation of extensive intercellular coupling and large numbers of gap junctions in the CNS suggests a syncytium-like organization of glial compartments.[59] Gap junctions form intercellular links between astrocytes,[60–62] neurons,[59,62,63] astrocytes-oligoden-drocytes,[64–66] astrocytes-neurons,[67–70] and more recently microglia.[71–73] Inflammation is a hallmark of various CNS diseases such as bacterial and viral infections, multiple sclerosis (MS), Alzheimer's disease, and cerebral ischemia. Changes in gap junction intercellular communication as reflected by alterations in Cx isoform expression have been associated with several of these CNS inflammatory diseases, including experimental autoimmune encephalomyelitis (EAE), an animal model for MS, Alzheimer's disease, and cerebral ischemia,[74–79] which may have dramatic implications for the survival of neuronal and glial populations in the context of neuroinflammation.

One CNS infectious disease in which nothing is known regarding its impact on glial gap junction communication (GJC) is parenchymal infection with pyogenic bacteria leading to the establishment of brain abscess. The experimental brain abscess model provides an excellent model system to study the consequences of neuroinflammation on GJC. Numerous proinflammatory mediators produced during the course of brain abscess development, including IL-1, TNF-α, and NO, have been reported to inhibit GJC in astrocytes[80–83] and, by extension, may be capable of modulating the type and extent of intercellular coupling in the context of brain abscess. In addition to the anticipated local changes in gap junction expression within the developing abscess, the massive proinflammatory milieu that ensues following infection is expected to impact the surrounding glial syncytial network, which may have long-reaching implications on neuron homeostasis and viability within brain regions physically removed from the primary abscess. Moreover, astrocytic GJC with endothelial cells seems to be involved in BBB maintenance,[84–87,88] and BBB dysfunction is a highly deleterious event in brain abscesses as it contributes to tissue necrosis.[13,14]

We have already established a link between TLR2 engagement and proinflammatory mediator production in primary astrocytes triggered by *S. aureus* treatment.[24] In addition, recent studies from our laboratory have demonstrated that both *S. aureus* and PGN inhibit GJC in astrocytes.[25] Ongoing studies in the laboratory are investigating the functional importance of TLR2- and MyD88-dependent pathways in regulating alterations in astrocytic coupling in response to *S. aureus*. A technique to study functional GJC in primary glia that can be used to assess the effects of a wide array of inflammatory stimuli on glial communication is described below.

9.3.2 EVALUATING FUNCTIONAL GAP JUNCTION CHANNELS IN GLIA

Traditionally, GJC has been demonstrated experimentally by tracking the transfer of gap junction–permeable dyes, such as Lucifer yellow (LY; 453 Da), to neighboring coupled cells. This can be accomplished by scrape loading,[83,89,90] preloading dye techniques,[91] or a more precise method of intracellular injections to delineate the degree of dye transfer.[25,71] The following protocol describes the steps required for intracellular microinjections of LY to evaluate the extent of glial GJC in vitro.[25]

Protocol 9.3: Single-Cell Microinjection Technique with Lucifer Yellow to Demonstrate Functional Gap Junction Channels in Astrocytes

1. Astrocytes (5×10^5 cells) are plated in 35 mm dishes and incubated for 24 hours. When working with alternative cell types, the seeding density will likely differ. Longer incubation periods can be used, as it is generally accepted that maximal gap junction formation requires time, and some studies report culturing cells for a period of up to 2 weeks.[92] After this resting period following cell seeding, astrocytes are exposed to the particular stimulus of interest and incubated for the desired time interval. We routinely examine the effects of *S. aureus* stimulation on astrocyte GJC between 12 and 24 hours following bacterial exposure (e.g., 10^7 colony-forming units per well, or 10 µg/ml of PGN).

2. Cultures are perfused with carbogen (95% O_2 and 5% CO_2) saturated Ames medium (Sigma-Aldrich, St. Louis, MO) at a rate of 5 ml/minute. The perfusion solution is prewarmed to 37 °C using an inline solution heater regulated by a dual channel heater (both from Warner Instruments, Hamden, CT). Cells are allowed to equilibrate for 15 minutes prior to microinjections to prevent any artifacts from the change in bathing medium on cell coupling.

3. Sharp micropipettes (resistance 100–200 MΩ) are prepared from borosilicate filament tubing (outer diameter [OD], 1.0 mm; inner diameter [ID], 0.5 mm) using a micropipette puller (P-97 Fleming/Brown; both from Sutter Instruments, Novato, CA). Pipettes are backfilled with the gap junction–permeable dye Lucifer yellow CH (4% w/v [i.e., by weight] LY in 150 mM LiCl; Sigma-Aldrich, St. Louis, MO) with a 34 gauge microfill syringe needle (World Precision Instruments, Sarasota, FL). LY is injected intracellularly into a single astrocyte by applying brief hyperpolarizing pulses delivered by a microiontophoresis current generator (World Precision Instruments) through a micropipette that is carefully inserted into the cell guided by a micromanipulator (MP-225; Sutter Instruments). This approach leads to the exit of LY from the micropipette based on the repulsion of the negatively charged LY by the negative current. A minimum of 10 individual cells are microinjected with LY in each experimental group to ensure that an adequate sample size is obtained for statistical analysis. We routinely perform intracellular injections in a minimum of three independent studies to allow the number of coupled cells in response to treatment to be compiled across experiments.

4. LY-injected cells are visualized using a fixed-stage upright epifluorescence microscope (BX51WI; Olympus, New Hyde Park, NY) equipped with a 40× water immersion objective lens (numerical aperture [N.A.] 0.8), a 12-bit intensified monochrome CCD camera (CoolSnap ES; Photometrics, Tucson, AZ), a multiple excitation filter controlled through a filter wheel (Prior Scientific, Rockland, MA), and a quad emission filter (Chroma, Rockingham, VT). Images are acquired and processed using MetaMorph software

(Universal Imaging, Downingtown, PA). Images depicting the extent of LY spread are presented after background subtraction for LY (excitation wave length 425 nm, and emission wave length 540 nm) to correct for autofluorescence. In terms of analysis, the number of gap junction–coupled cells can be enumerated and/or the distance of dye spread from the original microinjected cell can be measured.

In studies designed to confirm the gap junction–dependent nature of LY spread in astrocytes, cells can be treated with any number of gap junction blockers, including 18-α-glycyrrhetinic acid (AGA), carbenoxolone, octanol, or heptanol.[93] As a control, chemically related but inactive compounds such as glycyrrhizic acid (GZA) can be used to demonstrate specificity in GJC blockade.[93] For blockade studies, a minimum of three LY intracellular injections should be performed in unstimulated astrocyte cultures prior to gap junction inhibitor treatment to confirm the establishment of gap junction–dependent coupling. Subsequently, the same cultures are continuously perfused for 20 minutes in Ames medium supplemented with the gap junction blocker to allow for sufficient levels of inhibitor to accumulate within the culture medium, whereupon gap junction–dependent coupling can be assessed while still in the presence of the blocking agent. Although these blockers are used frequently by investigators in the gap junction field, unfortunately, these compounds are not specific inhibitors of gap junctions but also affect various subsets of ion channels.[93] Alternative strategies that can be employed to more specifically attenuate gap junction activity include siRNA for a particular Cx isoform or Cx mimetic peptides;[94,95] however, one must first determine the major Cx isoform composition of the gap junction channels in a particular cell type. In addition, the types of Cx isoforms constituting gap junction channels can be quite complex, and an approach to inhibit multiple Cxs at once (i.e., via siRNA) may lead to more overt phenotypes since Cx isoforms have been reported to exhibit functional redundancy.[59,96,97]

Gap junction communication is highly regulated because it plays an essential role in the maintenance of proper CNS homeostasis. Connexin proteins have a short half-life of approximately 2–5 hours in culture, and thus their expression is tightly controlled at multiple levels, including transcriptional regulation of Cx gene expression, translational control of Cx expression, as well as posttranslational modifications.[58,98,99] Moreover, gap junction channels are tightly regulated through assembly, such as the modulation of hemichannel trafficking to the cell membrane, docking, and the formation of gap junction plaques.[56,58] In addition, the activity of gap junction channels can be influenced through variations in pH, proteasome, and/or lysosomal degradation pathways for Cx proteins and the phosphorylation state of Cx isoforms.[58,98,99]

Another level at which GJC is regulated is through phosphorylation of Cx isoforms that dictates the open-closed nature of gap junction channels. Connexin 43 (Cx43) is a phosphoprotein that can exist in a nonphosphorylated form and two phosphorylated species.[56,100] Connexin proteins are primarily phosphorylated on serine residues with some residual phosphothreonine and phosphotyrosine.[100] The phosphorylation state of Cx43 is thought to be highly regulated via multiple pathways, including mitogen-activated protein kinase (MAPK), Src, and protein kinase

C (PKC).[56,101] Specifically, studies have suggested that serine kinases activated by MAPK- and PKC-dependent pathways phosphorylate Cx43-containing gap junction channels, closing these intercellular conduits.[102] The identified phosphorylation sites for Cx43 by PKC are ser-368 and ser-372,[103] whereas ser-255, ser-279, and ser-282 are targeted for phosphorylation by MAPK.[101,104] We have recently demonstrated that the p38 MAPK pathway plays an important role in controlling the resultant block in gap junction communication observed in primary astrocytes in response to the gram-positive pathogen *S. aureus* as well as the TLR2 ligand PGN.[25] Therefore, the use of inhibitors for various signal transduction pathways represents a tool to investigate the downstream signaling events originating from TLR engagement and how this affects glial gap junction communication.

ACKNOWLEDGMENTS

I would like to thank Dr. Nilufer Esen for critical review of the manuscript. Dr. Kielian's laboratory is supported by grants from the National Institute of Neurological Disorders and Stroke (NINDS; RO1s NS053487, NS40730, and NS055385). Additional support for Dr. Kielian's research is provided by the NINDS-supported core facility at University of Arkansas for Medical Sciences (UAMS; P30 NS047546).

REFERENCES

1. Nau, R., and Bruck, W. Neuronal injury in bacterial meningitis: mechanisms and implications for therapy. Trends Neurosci 25 (1), 38–45, 2002.
2. Koedel, U., Scheld, W. M., and Pfister, H. W. Pathogenesis and pathophysiology of pneumococcal meningitis. Lancet Infect Dis 2 (12), 721–36, 2002.
3. Merkelbach, S., Sittinger, H , Schweizer, I., and Muller, M. Cognitive outcome after bacterial meningitis. Acta Neurol Scand 102 (2), 118–23, 2000.
4. Grimwood, K., Anderson, P., Anderson, V., Tan, L., and Nolan, T. Twelve year outcomes following bacterial meningitis: further evidence for persisting effects. Arch Dis Child 83 (2), 111–16, 2000.
5. Lehnardt, S., Massillon, L., Follett, P., Jensen, F. E., Ratan, R., Rosenberg, P. A., Volpe, J. J., and Vartanian, T. Activation of innate immunity in the CNS triggers neurodegeneration through a Toll-like receptor 4-dependent pathway. Proc Natl Acad Sci USA 100 (14), 8514–9, 2003.
6. Lehnardt, S., Lachance, C., Patrizi, S., Lefebvre, S., Follett, P. L., Jensen, F. E., Rosenberg, P. A., Volpe, J. J., and Vartanian, T. The toll-like receptor TLR4 is necessary for lipopolysaccharide-induced oligodendrocyte injury in the CNS. J Neurosci 22 (7), 2478–86, 2002.
7. Echchannaoui, H., Frei, K., Schnell, C., Leib, S. L., Zimmerli, W., and Landmann, R. Toll-like receptor 2-deficient mice are highly susceptible to Streptococcus pneumoniae meningitis because of reduced bacterial clearing and enhanced inflammation. J Infect Dis 186 (6), 798–806, 2002.
8. Koedel, U., Angele, B., Rupprecht, T., Wagner, H., Roggenkamp, A., Pfister, H. W., and Kirschning, C. J. Toll-like receptor 2 participates in mediation of immune response in experimental pneumococcal meningitis. J Immunol 170 (1), 438–44, 2003.
9. Koedel, U., Rupprecht, T., Angele, B., Heesemann, J., Wagner, H., Pfister, H. W., and Kirschning, C. J. MyD88 is required for mounting a robust host immune response to Streptococcus pneumoniae in the CNS. Brain 127 (Pt. 6), 1437–45, 2004.

10. Kielian, T. Immunopathogenesis of brain abscess. J Neuroinflammation 1 (1), 16, 2004.

11. Mathisen, G. E., and Johnson, J. P. Brain abscess. Clin Infect Dis 25 (4), 763–79, quiz 780–81, 1997.

12. Townsend, G. C., and Scheld, W. M. Infections of the central nervous system. Adv Intern Med 43, 403–47, 1998.

13. Bloch, O., Papadopoulos, M. C., Manley, G. T., and Verkman, A. S. Aquaporin-4 gene deletion in mice increases focal edema associated with staphylococcal brain abscess. J Neurochem 95 (1), 254–62, 2005.

14. Lo, W. D., Wolny, A., and Boesel, C. Blood-brain barrier permeability in staphylococcal cerebritis and early brain abscess. J Neurosurg 80 (5), 897–905, 1994.

15. Lu, C. H., Chang, W. N., and Lui, C. C. Strategies for the management of bacterial brain abscess. J Clin Neurosci 13 (10), 979–85, 2006.

16. Stenzel, W., Soltek, S., Miletic, H., Hermann, M. M., Korner, H., Sedgwick, J. D., Schluter, D., and Deckert, M. An essential role for tumor necrosis factor in the formation of experimental murine Staphylococcus aureus-induced brain abscess and clearance. J Neuropathol Exp Neurol 64 (1), 27–36, 2005.

17. Kielian, T., Phulwani, N. K., Esen, N., Syed, M. M., Haney, A. C., McCastlain, K., and Johnson, J. MyD88-dependent signals are essential for the host immune response in experimental brain abscess. J Immunol 178 (7), 4528–37, 2007.

18. Kielian, T., and Hickey, W. F. Proinflammatory cytokine, chemokine, and cellular adhesion molecule expression during the acute phase of experimental brain abscess development. Am J Pathol 157 (2), 647–58, 2000.

19. Baldwin, A. C., and Kielian, T. Persistent immune activation associated with a mouse model of Staphylococcus aureus-induced experimental brain abscess. J Neuroimmunol 151 (1–2), 24–32, 2004.

20. Kielian, T., Barry, B., and Hickey, W. F. CXC chemokine receptor-2 ligands are required for neutrophil-mediated host defense in experimental brain abscesses. J Immunol 166 (7), 4634–43, 2001.

21. Flaris, N. A., and Hickey, W. F. Development and characterization of an experimental model of brain abscess in the rat. Am J Pathol 141 (6), 1299–307, 1992.

22. Stenzel, W., Soltek, S., Schluter, D., and Deckert, M. The intermediate filament GFAP is important for the control of experimental murine Staphylococcus aureus-induced brain abscess and Toxoplasma encephalitis. J Neuropathol Exp Neurol 63 (6), 631–40, 2004.

23. Kimelberg, H. K. Water homeostasis in the brain: basic concepts. Neuroscience 129 (4), 851–60, 2004.

24. Esen, N., Tanga, F. Y., DeLeo, J. A., and Kielian, T. Toll-like receptor 2 (TLR2) mediates astrocyte activation in response to the Gram-positive bacterium Staphylococcus aureus. J Neurochem 88 (3), 746–58, 2004.

25. Esen, N., Shuffield, D., Syed, M. M., and Kielian, T. Modulation of connexin expression and gap junction communication in astrocytes by the gram-positive bacterium S. aureus. Glia 55 (1), 104–17, 2007.

26. Kielian, T., Esen, N., and Bearden, E. D. Toll-like receptor 2 (TLR2) is pivotal for recognition of S. aureus peptidoglycan but not intact bacteria by microglia. Glia 49 (4), 567–76, 2005.

27. Kielian, T., Mayes, P., and Kielian, M. Characterization of microglial responses to Staphylococcus aureus: effects on cytokine, costimulatory molecule, and Toll-like receptor expression. J Neuroimmunol 130 (1–2), 86–99, 2002.

28. Hoffmann, O., Braun, J. S., Becker, D., Halle, A., Freyer, D., Dagand, E., Lehnardt, S., and Weber, J. R. TLR2 mediates neuroinflammation and neuronal damage. J Immunol 178 (10), 6476–81, 2007.

29. Kielian, T. Toll-like receptors in central nervous system glial inflammation and homeostasis. J Neurosci Res 83 (5), 711–30, 2006.
30. Olson, J. K., and Miller, S. D. Microglia initiate central nervous system innate and adaptive immune responses through multiple TLRs. J Immunol 173 (6), 3916–24, 2004.
31. Carpentier, P. A., Begolka, W. S., Olson, J. K., Elhofy, A., Karpus, W. J., and Miller, S. D. Differential activation of astrocytes by innate and adaptive immune stimuli. Glia 49 (3), 360–74, 2005.
32. Bsibsi, M., Ravid R., Gveric D., and van Noort J. M. Broad expression of Toll-like receptors in the human central nervous system.. J Neuropathol Exp Neurol 61 (11), 1013–21, 2002.
33. Bowman, C. C., Rasley, A., Tranguch, S. L., and Marriott, I. Cultured astrocytes express toll-like receptors for bacterial products. Glia 43 (3), 281–91, 2003.
34. Babcock, A. A., Wirenfeldt, M., Holm, T., Nielsen, H. H., Dissing-Olesen, L., Toft-Hansen, H., Millward, J. M., Landmann, R., Rivest, S., Finsen, B., and Owens, T. Toll-like receptor 2 signaling in response to brain injury: an innate bridge to neuroinflammation. J Neurosci 26 (49), 12826–37, 2006.
35. Tanga, F. Y., Raghavendra, V., and DeLeo, J. A. Quantitative real-time RT-PCR assessment of spinal microglial and astrocytic activation markers in a rat model of neuropathic pain. Neurochem Int 45 (2–3), 397–407, 2004.
36. Kigerl, K. A., Lai, W., Rivest, S., Hart, R. P., Satoskar, A. R., and Popovich, P. G. Toll-like receptors (TLR)-2 and TLR-4 regulate inflammation, gliosis, and myelin sparing after spinal cord injury. J Neurochem 102 (1), 37-50, 2007.
37. Esen, N., and Kielian, T. Central role for MyD88 in the responses of microglia to pathogen-associated molecular patterns. J Immunol 176 (11), 6802–11, 2006.
38. Block, M. L., Zecca, L., and Hong, J. S. Microglia-mediated neurotoxicity: uncovering the molecular mechanisms. Nat Rev Neurosci 8 (1), 57–69, 2007.
39. Qin, L., Li, G., Qian, X., Liu, Y., Wu, X., Liu, B., Hong, J. S., and Block, M. L. Interactive role of the toll-like receptor 4 and reactive oxygen species in LPS-induced microglia activation. Glia 52 (1), 78–84, 2005.
40. Jack, C. S., Arbour, N., Manusow, J., Montgrain, V., Blain, M., McCrea, E., Shapiro, A., and Antel, J. P. TLR signaling tailors innate immune responses in human microglia and astrocytes. J Immunol 175 (7), 4320–30, 2005.
41. Lotz, M., Ebert, S., Esselmann, H., Iliev, A. I., Prinz, M., Wiazewicz, N., Wiltfang, J., Gerber, J., and Nau, R. Amyloid beta peptide 1-40 enhances the action of Toll-like receptor-2 and -4 agonists but antagonizes Toll-like receptor-9-induced inflammation in primary mouse microglial cell cultures. J Neurochem 94 (2), 289–98, 2005.
42. Goos, M., Lange, P., Hanish, U-K., Prinz, M., Scheffel, J., Bergmann, R., Ebert, S., and Nau, R. Fibronectin is elevated in the cerebrospinal fluid of patients suffering from bacterial meningitis and enhances inflammation caused by bacterial products in primary mouse microglial cell cultures. Journal of Neurochemistry 102 (6), 2049–60, 2007.
43. Tsan, M. F., and Gao, B. Endogenous ligands of Toll-like receptors. J Leukoc Biol 76 (3), 514–19, 2004.
44. Herrmann, I., Kellert, M., Schmidt, H., Mildner, A., Hanisch, U. K., Bruck, W., Prinz, M., and Nau, R. Streptococcus pneumoniae infection aggravates experimental autoimmune encephalomyelitis via Toll-like receptor 2. Infect Immun 74 (8), 4841–8, 2006.
45. Lehnardt, S., Henneke, P., Lien, E., Kasper, D. L., Volpe, J. J., Bechmann, I., Nitsch, R., Weber, J. R., Golenbock, D. T., and Vartanian, T. A mechanism for neurodegeneration induced by group B streptococci through activation of the TLR2/MyD88 pathway in microglia. J Immunol 177 (1), 583–92, 2006.
46. Tanga, F. Y., Nutile-McMenemy, N., and DeLeo, J. A. The CNS role of Toll-like receptor 4 in innate neuroimmunity and painful neuropathy. Proc Natl Acad Sci USA 102 (16), 5856–61, 2005.

47. Borgmann, K., Gendelman, H. E., and Ghorpade, A. Isolation and HIV-1 infection of primary human microglia from fetal and adult tissue. Methods Mol Biol 304, 49–70, 2005.

48. Hassan, N. F., Campbell, D. E., Rifat, S., and Douglas, S. D. Isolation and characterization of human fetal brain-derived microglia in in vitro culture. Neuroscience 41 (1), 149–58, 1991.

49. Esen, N., and Kielian, T. Effects of low dose GM-CSF on microglial inflammatory profiles to diverse pathogen-associated molecular patterns (PAMPs). J Neuroinflammation 4, 10, 2007.

50. Giulian, D., and Ingeman, J. E. Colony-stimulating factors as promoters of ameboid microglia. J Neurosci 8 (12), 4707–17, 1988.

51. Carson, M. J., Reilly, C. R., Sutcliffe, J. G., and Lo, D. Mature microglia resemble immature antigen-presenting cells. Glia 22 (1), 72–85, 1998.

52. Floden, A. M., and Combs, C. K. Beta-amyloid stimulates murine postnatal and adult microglia cultures in a unique manner. J Neurosci 26 (17), 4644–8, 2006.

53. Ponomarev, E. D., Novikova, M., Maresz, K., Shriver, L. P., and Dittel, B. N. Development of a culture system that supports adult microglial cell proliferation and maintenance in the resting state. J Immunol Methods 300 (1–2), 32–46, 2005.

54. Remahl, S., and Hildebrand, C. Changing relation between onset of myelination and axon diameter range in developing feline white matter. J Neurol Sci 54 (1), 33–45, 1982.

55. Macklin, W. B., and Weill, C. L. Appearance of myelin proteins during development in the chick central nervous system. Dev Neurosci 7 (3), 170–78, 1985.

56. Saez, J. C., Berthoud, V. M., Branes, M. C., Martinez, A. D., and Beyer, E. C. Plasma membrane channels formed by connexins: their regulation and functions. Physiol Rev 83 (4), 1359–400, 2003.

57. Spray, D. C., Duffy, H. S., and Scemes, E. Gap junctions in glia: types, roles, and plasticity. In The functional roles of glial cells in health and disease, ed. Tsacopoulos, M. A. New York: Plenum, 1999, pp. 339–59.

58. Laird, D. W. Life cycle of connexins in health and disease. Biochem J 394 (Pt. 3), 527–43, 2006.

59. Rouach, N., Avignone, E., Meme, W., Koulakoff, A., Venance, L., Blomstrand, F., and Giaume, C. Gap junctions and connexin expression in the normal and pathological central nervous system. Biol Cell 94 (7–8), 457–75, 2002.

60. Binmoller, F. J., and Muller, C. M. Postnatal development of dye-coupling among astrocytes in rat visual cortex. Glia 6 (2), 127–37, 1992.

61. D'Ambrosio, R., Wenzel, J., Schwartzkroin, P. A., McKhann, G. M., Jr., and Janigro, D. Functional specialization and topographic segregation of hippocampal astrocytes. J Neurosci 18 (12), 4425–38, 1998.

62. Dermietzel, R. Molecular diversity and plasticity of gap junctions in the nervous system. In Gap junctions in the nervous system, ed. Spray, D. C., and Dermietzel, R. New York: Chapman & Hall, 1996, pp. 13–38.

63. Rozental, R., Giaume, C., and Spray, D. C. Gap junctions in the nervous system. Brain Res Rev 32 (1), 11–15, 2000.

64. Massa, P. T., and Mugnaini, E. Cell junctions and intramembrane particles of astrocytes and oligodendrocytes: a freeze-fracture study. Neuroscience 7 (2), 523–38, 1982.

65. Massa, P. T., and Mugnaini, E. Cell-cell junctional interactions and characteristic plasma membrane features of cultured rat glial cells. Neuroscience 14 (2), 695–709, 1985.

66. Rash, J. E., Duffy, H. S., Dudek, F. E., Bilhartz, B. L., Whalen, L. R., and Yasumura, T. Grid-mapped freeze-fracture analysis of gap junctions in gray and white matter of adult rat central nervous system, with evidence for a "panglial syncytium" that is not coupled to neurons. J Comp Neurol 388 (2), 265–92, 1997.

67. Alvarez-Maubecin, V., Garcia-Hernandez, F., Williams, J. T., and Van Bockstaele, E. J. Functional coupling between neurons and glia. J Neurosci 20 (11), 4091–8, 2000.

68. Froes, M. M., Correia, A. H., Garcia-Abreu, J., Spray, D. C., Campos de Carvalho, A. C., and Neto, M. V. Gap-junctional coupling between neurons and astrocytes in primary central nervous system cultures. Proc Natl Acad Sci USA 96 (13), 7541–6, 1999.

69. Froes, M. M., and de Carvalho, A. C. Gap junction-mediated loops of neuronal-glial interactions. Glia 24 (1), 97–107, 1998.

70. Rash, J. E., Yasumura, T., Dudek, F. E., and Nagy, J. I. Cell-specific expression of connexins and evidence of restricted gap junctional coupling between glial cells and between neurons. J Neurosci 21 (6), 1983–2000, 2001.

71. Eugenin, E. A., Eckardt, D., Theis, M., Willecke, K., Bennett, M. V., and Saez, J. C. Microglia at brain stab wounds express connexin 43 and in vitro form functional gap junctions after treatment with interferon-gamma and tumor necrosis factor-alpha. Proc Natl Acad Sci USA 98 (7), 4190–5, 2001.

72. Martinez, A. D., Eugenin, E. A., Branes, M. C., Bennett, M. V., and Saez, J. C. Identification of second messengers that induce expression of functional gap junctions in microglia cultured from newborn rats. Brain Res 943 (2), 191–201, 2002.

73. Garg, S., Syed, M. M., and Kielian, T. Staphylococcus aureus-derived peptidoglycan induces Cx43 expression and functional gap junction intercellular communication in microglia. J Neurochem 95 (2), 475–83, 2005.

74. Nagy, J. I., Li, W., Hertzberg, E. L., and Marotta, C. A. Elevated connexin43 immunoreactivity at sites of amyloid plaques in Alzheimer's disease. Brain Res 717 (1–2), 173–78, 1996.

75. Siushansian, R., Bechberger, J. F., Cechetto, D. F., Hachinski, V. C., and Naus, C. C. Connexin43 null mutation increases infarct size after stroke. J Comp Neurol 440 (4), 387–94, 2001

76. Nakase, T., Fushiki, S., and Naus, C. C. Astrocytic gap junctions composed of connexin 43 reduce apoptotic neuronal damage in cerebral ischemia. Stroke 34 (8), 1987–93, 2003.

77. Nakase, T., Sohl, G., Theis, M., Willecke, K., and Naus, C. C. Increased apoptosis and inflammation after focal brain ischemia in mice lacking connexin43 in astrocytes. Am J Pathol 164 (6), 2067–75, 2004.

78. Brand-Schieber, E., Werner, P., Iacobas, D. A., Iacobas, S., Beelitz, M., Lowery, S. L., Spray, D. C., and Scemes, E. Connexin43, the major gap junction protein of astrocytes, is down-regulated in inflamed white matter in an animal model of multiple sclerosis. J Neurosci Res 80 (6), 798–808, 2005.

79. Roscoe, W. A., Messersmith, E., Meyer-Franke, A., Wipke, B., and Karlik, S. J. Connexin 43 gap junction proteins are up-regulated in remyelinating spinal cord. J Neurosci Res 85 (5), 945–53, 2007.

80. Meme, W., Ezan, P., Venance, L., Glowinski, J., and Giaume, C. ATP-induced inhibition of gap junctional communication is enhanced by interleukin-1 beta treatment in cultured astrocytes. Neuroscience 126 (1), 95–104, 2004.

81. John, G. R., Scemes, E., Suadicani, S. O., Liu, J. S., Charles, P. C., Lee, S. C., Spray, D. C., and Brosnan, C. F. IL-1beta differentially regulates calcium wave propagation between primary human fetal astrocytes via pathways involving P2 receptors and gap junction channels. Proc Natl Acad Sci USA 96 (20), 11613–8, 1999.

82. Zvalova, D., Cordier, J., Mesnil, M., Junier, M. P., and Chneiweiss, H. p38/SAPK2 controls gap junction closure in astrocytes. Glia 46 (3), 323–33, 2004.

83. Bolanos, J. P., and Medina, J. M. Induction of nitric oxide synthase inhibits gap junction permeability in cultured rat astrocytes. J Neurochem 66 (5), 2091–9, 1996.
84. Nagasawa, K., Chiba, H., Fujita, H., Kojima, T., Saito, T., Endo, T., and Sawada, N. Possible involvement of gap junctions in the barrier function of tight junctions of brain and lung endothelial cells. J Cell Physiol 208 (1), 123–32, 2006.
85. Hayashi, K., Nakao, S., Nakaoke, R., Nakagawa, S., Kitagawa, N., and Niwa, M. Effects of hypoxia on endothelial/pericytic co-culture model of the blood-brain barrier. Regul Pept 123 (1–3), 77–83, 2004.
86. Paemeleire, K. Calcium signaling in and between brain astrocytes and endothelial cells. Acta Neurol Belg 102 (3), 137–40, 2002.
87. Leybaert, L., Cabooter, L., and Braet, K. Calcium signal communication between glial and vascular brain cells. Acta Neurol Belg 104 (2), 51–56, 2004.
88. Simard, M., Arcuino, G., Takano, T., Liu, Q. S., and Nedergaard, M. Signaling at the gliovascular interface. J Neurosci 23 (27), 9254–62, 2003.
89. el-Fouly, M. H., Trosko, J. E., and Chang, C. C. Scrape-loading and dye transfer: a rapid and simple technique to study gap junctional intercellular communication. Exp Cell Res 168 (2), 422–30, 1987.
90. Giaume, C., Marin, P., Cordier, J., Glowinski, J., and Premont, J. Adrenergic regulation of intercellular communications between cultured striatal astrocytes from the mouse. Proc Natl Acad Sci USA 88 (13), 5577–81, 1991.
91. Goldberg, G. S., Bechberger, J. F., and Naus, C. C. A pre-loading method of evaluating gap junctional communication by fluorescent dye transfer. Biotechniques 18 (3), 490–97, 1995.
92. Giaume, C., Fromaget, C., el Aoumari, A., Cordier, J., Glowinski, J., and Gros, D. Gap junctions in cultured astrocytes: single-channel currents and characterization of channel-forming protein. Neuron 6 (1), 133–43, 1991.
93. Rozental, R., Srinivas, M., and Spray D. C. How to close a gap junction channel: efficacies and potencies of uncoupling agents. In Methods in molecular biology, ed. Bruzzone, R. A., and Giaume, C. Totowa, NJ: Humana Press, 2000, pp. 447–76.
94. De Vuyst, E., Decrock, E., Cabooter, L., Dubyak, G. R., Naus, C. C., Evans, W. H., and Leybaert, L. Intracellular calcium changes trigger connexin 32 hemichannel opening. Embo J 25 (1), 34–44, 2006.
95. Martin, P. E., Wall, C., and Griffith, T. M. Effects of connexin-mimetic peptides on gap junction functionality and connexin expression in cultured vascular cells. Br J Pharmacol 144 (5), 617–27, 2005.
96. Dermietzel, R., Gao, Y., Scemes, E., Vieira, D., Urban, M., Kremer, M., Bennett, M. V., and Spray, D. C. Connexin43 null mice reveal that astrocytes express multiple connexins. Brain Res Rev 32 (1), 45–56, 2000.
97. Naus, C. C., Bechberger, J. F., Zhang, Y., Venance, L., Yamasaki, H., Juneja, S. C., Kidder, G. M., and Giaume, C. Altered gap junctional communication, intercellular signaling, and growth in cultured astrocytes deficient in connexin43. J Neurosci Res 49 (5), 528–40, 1997.
98. Solan, J. L., and Lampe, P. D. Connexin phosphorylation as a regulatory event linked to gap junction channel assembly. Biochim Biophys Acta 1711 (2), 154–63, 2005.
99. Berthoud, V. M., Minogue, P. J., Laing, J. G., and Beyer, E. C. Pathways for degradation of connexins and gap junctions. Cardiovasc Res 62 (2), 256–67, 2004.
100. Warn-Cramer, B. J., and Lau, A. F. Regulation of gap junctions by tyrosine protein kinases. Biochim Biophys Acta 1662 (1–2), 81–95, 2004.
101. Warn-Cramer, B. J., Cottrell, G. T., Burt, J. M., and Lau, A. F. Regulation of connexin-43 gap junctional intercellular communication by mitogen-activated protein kinase. J Biol Chem 273 (15), 9188–96, 1998.

102. Saez, J. C., Retamal, M. A., Basilio, D., Bukauskas, F. F., and Bennett, M. V. Connexin-based gap junction hemichannels: gating mechanisms. Biochim Biophys Acta 1711 (2), 215–24, 2005.
103. Saez, J. C., Nairn, A. C., Czernik, A. J., Fishman, G. I., Spray, D. C., and Hertzberg, E. L. Phosphorylation of connexin43 and the regulation of neonatal rat cardiac myocyte gap junctions. J Mol Cell Cardiol 29 (8), 2131–45, 1997.
104. Warn-Cramer, B. J., Lampe, P. D., Kurata, W. E., Kanemitsu, M. Y., Loo, L. W., Eckhart, W., and Lau, A. F. Characterization of the mitogen-activated protein kinase phosphorylation sites on the connexin-43 gap junction protein. J Biol Chem 271 (7), 3779–86, 1996.

10 The Influence of Injury on Toll-Like Receptor Responses

Thomas J. Murphy, Adrian A. Maung,
Hugh M. Paterson, and James A. Lederer

CONTENTS

10.1 INTRODUCTION AND OVERVIEW

Toll-like receptors (TLRs) are a family of evolutionarily conserved cell-surface and intracellular receptors that react to bacterial or viral antigens or to endogenous factors released during cell injury. The ability to recognize a variety of common microbial antigens and endogenous factors indicates that a primary function of TLRs is to act as sentinel receptors to alert the innate immune system to infection or tissue damage. Once triggered, TLRs initiate strong inflammatory responses and set in motion innate and adaptive immune responses with the intent to help the infected or injured host combat potentially harmful infections. Because of their established importance in signaling infections, the microbial ligands for TLRs have been studied more extensively than the endogenous stress- or injury-induced TLR ligands. Nevertheless, the interplay between tissue injury and TLR responses could potentially be as important as their role in signaling microbial responses. This chapter will discuss recent advances in our understanding of the role TLRs play in signaling or regulating the injury response.

10.1.1 Toll-Like Receptors

Toll was first identified in the fruit fly, *Drosophila melanogaster*, as a gene important for dorsoventral development.[1,2] Subsequent pioneering work with Toll mutant flies demonstrated that flies lacking Toll were highly susceptible to fungal infection.[3] This suggested that Toll plays a central role in protective immunity in fruit flies. Toll receptor homologues with similar immune function were eventually identified in plants, insects, worms, and vertebrate mammals, indicating that Toll receptors are evolutionarily conserved immune-signaling receptors.[4] Toll-like receptors are the mammalian homologues of *Drosophila* Toll. To date, 13 mammalian TLRs have been identified: TLRs 1–10 in humans and TLRs 1–13 minus TLR10 in mice.[5] All TLRs are characterized as type I transmembrane receptors with an extracellular leucine-rich repeat (LRR) domain and a cytoplasmic tail with high similarity to the type 1 interleukin-1 (IL-1) receptor.[6] Appropriately, this shared signaling motif is called the Toll/IL-1 receptor (TIR) domain. The LRR domains of TLRs bind different microbial components, often referred to as pathogen-associated molecular patterns (PAMPs).[7] Because TLRs recognize PAMPs, they are appropriately called pattern recognition receptors (PRRs). Examples of PAMPs include bacteria cell wall molecules such as lipopolysaccharide (LPS), bacterial lipopeptide (BLP), peptidoglycan (PGN), and lipoteichoic acid (LTA). Interestingly, some TLRs are expressed on the cell surface, while others are expressed inside cells on the membranes of endocytic vesicles.

Many ligands for different TLRs have been identified and are listed in Table 10.1. Some TLRs bind several different PAMPs, and other TLRs can combine as heterodimers (e.g., TLR2-TLR1 and TLR2-TLR6) to provide diversity for PAMP recognition. However, specific ligands for TLR10, TLR12, and TLR13 have not yet been identified. As listed in Table 10.1, TLRs that recognize microbial ligands also recognize endogenous ligands that are released by cell stress and injury.[8] These include several different heat shock proteins (HSPs), some extracellular matrix proteins, high-mobility group box 1 (HMGB1), and an antimicrobial peptide.[9–12] These endogenous TLR ligands are suspected as playing a role in initiating, amplifying, or regulating the immune response to tissue damage. As such, endogenous TLR ligands have been referred to as damage-associated molecular patterns (DAMPs) or alarmins.[13–16] A major controversy regarding the validity of these endogenous TLR ligands is that their biological activity is mediated by microbial contaminants associated with preparations of endogenous TLR ligands. This is especially a concern for bacterial LPS and lipopeptide contamination since most described endogenous TLR ligands appear to be TLR4 and TLR2 restricted. Bacterial LPS and lipopeptides are also ubiquitous and difficult to remove from recombinant or purified preparations of these putative endogenous TLR ligands. Thus, careful confirmation of endogenous TLR ligands will be needed in order to clarify their importance in controlling immune responses to stress and injury.

10.1.2 TLRs and Immune Function

Interest in TLRs as being important elements of innate immune function in mammals was initiated by the discovery that the *Lps* gene, which had been identified by

TABLE 10.1
Microbial and Endogenous TLR Ligands

TLR	Microbial Ligands	Endogenous Ligands
TLR1 and TLR2	Triacylated lipopeptides[a]	
TLR2	Peptidoglycan, lipoteichoic acid, lipoarabinomannan, and porins[b]	HSP70 and GP96[c,d]
TLR2 and TLR6	Diacylated lipopeptides[e]	
TLR3	dsRNA[f]	
TLR4	LPS[g]	HSP60, HSP70, GP96, hyaluronan, lung surfactant protein A, fibronectin, fibrinogen, heparan sulfate, and β-defensin 2[h]
TLR5	Flagellin[i]	
TLR7	Guanosine- or uridine-rich ssRNA[j,k]	
TLR8	Guanosine- or uridine-rich ssRNA[j,k]	
TLR9	Unmethylated CpG DNA[l]	Chromatin-IgG complexes[m]
TLR10	Unknown	
TLR11	*T. gondii profilin-like protein*[n]	
TLR12	Unknown	
TLR13	Unknown	

Sources: a. Takeuchi, O., Sato, S., Horiuchi, T., Hoshino, K., Takeda, K., Dong, Z., Modlin, R. L., and Akira, S. Cutting edge: role of Toll-like receptor 1 in mediating immune response to microbial lipoproteins. *J Immunol* 169 (1), 10–14, 2002. b. Miyake, K. Innate immune sensing of pathogens and danger signals by cell surface Toll-like receptors. *Semin Immunol* 19 (1), 3–10, 2007. c. Vabulas, R. M., Ahmad-Nejad, P., Ghose, S., Kirschning, C. J., Issels, R. D., and Wagner, H. HSP70 as endogenous stimulus of the Toll/interleukin-1 receptor signal pathway. *J Biol Chem* 277 (17), 15107–12, 2002. d. Vabulas, R. M., Braedel, S., Hilf, N., Singh-Jasuja, H., Herter, S., Ahmad-Nejad, P., Kirschning, C. J., Da Costa, C., Rammensee, H. G., Wagner, H., and Schild, H. The endoplasmic reticulum-resident heat shock protein Gp96 activates dendritic cells via the Toll-like receptor 2/4 pathway. *J Biol Chem* 277 (23), 20847–53, 2002. e. Buwitt-Beckmann, U., Heine, H., Wiesmuller, K. H., Jung, G., Brock, R., Akira, S., and Ulmer, A. J. TLR1- and TLR6-independent recognition of bacterial lipopeptides. *J Biol Chem* 281 (14), 9049–57, 2006. f. Alexopoulou, L., Holt, A. C., Medzhitov, R., and Flavell, R. A. Recognition of double-stranded RNA and activation of NF-kappaB by Toll-like receptor 3. *Nature* 413 (6857), 732–78, 2001. g. Poltorak, A., He, X., Smirnova, I., Liu, M. Y., Huffel, C. V., Du, X., Birdwell, D., Alejos, E., Silva, M., Galanos, C., Freudenberg, M., Ricciardi-Castagnoli, P., Layton, B., and Beutler, B. Defective LPS signaling in C3H/HeJ and C57BL/10ScCr mice: mutations in Tlr4 gene. *Science* 282 (5396), 2085–8, 1998. h. Beg, A. A. Endogenous ligands of Toll-like receptors: implications for regulating inflammatory and immune responses. *Trends Immunol* 23 (11), 509–12, 2002. i. Hayashi, F., Smith, K. D., Ozinsky, A., Hawn, T. R., Yi, E. C., Goodlett, D. R., Eng, J. K., Akira, S., Underhill, D. M., and Aderem, A. The innate immune response to bacterial flagellin is mediated by Toll-like receptor 5. *Nature* 410 (6832), 1099–103, 2001. j. Sioud, M. Single-stranded small interfering RNA are more immunostimulatory than their double-stranded counterparts: a central role for 2′-hydroxyl uridines in immune responses. *Eur J Immunol* 36 (5), 1222–30, 2006. k. Lee, J., Chuang, T. H., Redecke, V., She, L., Pitha, P. M., Carson, D. A., Raz, E., and Cottam, H. B. Molecular basis for the immunostimulatory activity of guanine nucleoside analogs: activation of Toll-like receptor 7. *Proc Natl Acad Sci USA* 8, 8, 2003. l. Hemmi, H., Takeuchi, O., Kawai, T., Kaisho, T., Sato, S., Sanjo, H., Matsumoto, M., Hoshino, K., Wagner, H., Takeda, K., and Akira, S. A Toll-like receptor recognizes bacterial DNA. *Nature* 408 (6813), 740–45, 2000. m. Leadbetter, E. A., Rifkin, I. R., Hohlbaum, A. M., Beaudette, B. C., Shlomchik, M. J., and Marshak-Rothstein, A. Chromatin-IgG complexes activate B cells by dual engagement of IgM and Toll-like receptors. *Nature* 416 (6881), 603–7, 2002. n. Yarovinsky, F., Zhang, D., Andersen, J. F., Bannenberg, G. L., Serhan, C. N., Hayden, M. S., Hieny, S., Sutterwala, F. S., Flavell, R. A., Ghosh, S., and Sher, A. TLR11 activation of dendritic cells by a protozoan profilin-like protein. *Science* 308 (5728), 1626–9, 2005.

gene mapping as being responsible for the resistance to lethal endotoxin or bacterial LPS challenge in mice, encoded TLR4.[17] Mice with natural mutations in the *Tlr4* gene did not respond to *E. coli* LPS or Lipid A, the portion of the LPS molecule responsible for its biological activity. It was subsequently shown that cells from genetically engineered *Tlr4* gene–deficient mice did not respond to LPS stimulation, confirming that TLR4 was indeed the signaling receptor for LPS.[18] Medzhitov and Janeway subsequently showed that TLR4 provides an inflammatory adjuvant signal needed to evoke CD4 T cell responses, suggesting that TLRs also play an important role in regulating adaptive immune responses.[19]

While this early work focused primarily on TLR4 and innate immune cell responses, the rapid discovery of TLR family members revealed that TLRs have specialized functions depending on their cellular distribution. It is now known that many different cell types express TLRs. However, it is clear that they are more widely expressed on innate immune cell types, including monocytes, macrophages, neutrophils, and dendritic cells. Their primary function on these immune cells is to act as PRRs for microbial PAMPs in the infected host. Once triggered, TLR signaling leads to the rapid production of proinflammatory cytokines, including those needed for effective control of bacterial and viral infections such as type 1 and 2 interferons (IFNs).[20,21] Another well-described function for TLRs is their ability to transform resting innate cell types, like dendritic cells, into potent antigen-presenting cells (APCs). This occurs via upregulation of co-stimulatory receptors such as CD80, CD86, and CD40 that are needed to promote CD4 and CD8 T cell activation.[22–27] In addition, TLR signaling induces the production of interleukin-12 (IL-12), IL-6, IL-10, transforming growth factor-β (TGFβ), IFN-α, and IFN-β by APCs, which help guide the differentiation of naïve CD4 T cells into T helper cell subsets.[28–31] Thus, TLR signaling promotes a rapid inflammatory response, which then leads to the development of effective adaptive immune system reactivity.

10.1.3   TLR Signaling

TLRs signal cellular responses (with the exception of TLR3) through the adaptor molecule myeloid differentiation factor 88 (MyD88).[32] TLR3 signals through a MyD88-independent pathway that is dependent on the TIR domain–containing adaptor inducing interferon-β (TRIF)[28,30] (also call TICAM1). TLR4 signals through both the MyD88-dependent and -independent pathways.[33] An additional adaptor molecule called TIR domain–containing adaptor molecule (TIRAP), also called Mal, plays a role in TLR2 and TLR4 signaling, but not in signaling by the other TLR family members.[34,35] Thus, most TLRs share the MyD88 proximal signaling adaptor, but other adaptor molecules—TRIF and TIRAP—provide additional complexity for TLR3, TLR2, and TLR4 signaling. Figure 10.1 provides an overview of the factors involved in TLR signaling that are discussed in this section.

After early signaling events, two major pathways are activated depending on which TLR adaptor molecules are involved in generating the signal. For those TLRs that signal through the MyD88 pathway (TLR1, 2, 4, 5, and 6). MyD88 associates with the TIR domains of TLRs leading to homophilic or heterophilic aggregation.[36–41] The interleukin receptor (IL-R-) associated kinases (IRAK) IRAK-1 and IRAK-4

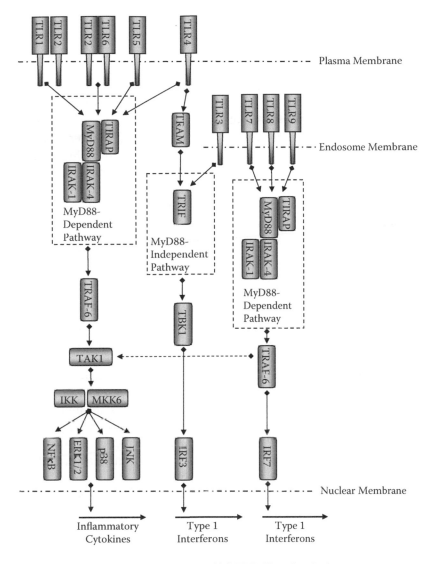

FIGURE 10.1 (Color figure follows page 104) TLR Signaling Pathways

are recruited to the complex and phosphorylated, after which IRAK-1 interacts with tumor necrosis factor (TNF) receptor-associated factor 6 (TRAF6). TRAF6 then associates with transforming growth factor-β–activating kinase 1 (TAK1), which then phosphorylates inhibitory kappa B kinase (IKK) and mitogen-activated protein (MAP) kinase kinase 6 (MKK6).[42-44] These signaling events cause the degradation of inhibitory kappa B (IκB) and the nuclear translocation of nuclear factor kappa B (NF-κB) as well as the phosphorylation of p38, ERK1/2, and JNK.[45] Thus, activation of the MyD88-dependent pathway induces inflammatory cytokines and a wide array of cellular responses associated with the development of inflammation and immune cell activation.

The MyD88-independent pathway can be triggered by a subset of TLRs. These include TLR3, TLR4, TLR7, TLR8, and TLR9.[30,46,47] While TLR4, 7, 8, and 9 can signal responses through both MyD88-dependent and MyD88-independent pathways, TLR3 has been shown to signal only through the MyD88-independent pathway. The MyD88-independent pathway plays an important role in antiviral immunity since the ligands for these TLRs are viral-derived molecules such as single- and double-stranded RNA (Table 10.1). In addition, these TLRs are located intracellularly within endosomes, suggesting that they signal responses to intracellular antigens such as those associated with viruses. Moreover, MyD88-independent TLR signaling induces type 1 IFN production, a potent antiviral cytokine. The signaling adaptor TRIF associates with TANK-binding kinase 1 (TBK1), which then phosphorylates IFN regulatory factor 3 (IRF3).[48] IRF3 then translocates into the nucleus to turn on IFN-β production. Once induced, IFN-β acts in a feedback manner to induce and amplify the production of other type 1 interferons by activating IRF7 and IRF8.[31]

Although a clearer picture of how TLRs signal cellular responses has emerged, there remains a number of unknown issues about TLR responses against microbial versus endogenous ligands. For example, most endogenous TLR ligands have been shown to stimulate TLR2 and TLR4 responses. What makes these two particular TLRs more reactive to endogenous ligands? In addition, are there endogenous ligands for other TLRs? For example, the discovery that TLR9 can signal self-DNA and self immune complex activation suggests that the regulation of self- and non-self-reactivity might involve the ability of cells to respond to self-antigens via TLRs or other related pattern recognition receptors.[49] Also, are there separate co-receptors responsible for microbial as opposed to endogenous TLR ligands? Some candidates for discriminating TLR2 and TLR4 reactivity include CD14, MD-1, and MD-2.[11] Our laboratory has addressed the significance of changes in TLR reactivity as being involved in boosting innate immune system responses after injury, whereas other groups have focused on the induction of tolerance to TLR responses following an infection as a feedback mechanism to control excessive inflammatory responses during severe infections like bacteremic sepsis.[50–52] In general, these physiological responses have a potent effect on TLR responses. In the case of injury, TLR responses are augmented as judged by an increase in inflammatory cytokine production by cells exposed to either TLR4 or TLR2 stimuli.[50] In contrast, sepsis causes a transient downregulation of TLR responses by a tolerance or tachyphylaxis mechanism.[53] In both of these responses, changes in TLR signaling have been shown to be involved in the altered TLR responses. The increase in TLR4 responsiveness following injury was attributable to increased phosphorylation of p38 in macrophages.[54] In contrast, the induction of LPS tolerance has been associated with decreased IRAK-1, elevated IRAK-M (a negative regulator of TLR4 signaling), and impaired recruitment of MyD88 due to impaired tyrosine phosphorylation of the TIR domain of TLR4.[51,55] Thus, TLR signaling is controlled by specific mechanisms that regulate the level of inflammation after injury or infection.

10.1.4 Injury Effects on TLR Responses

The findings from studies demonstrating that necrotic tissue can stimulate cells to produce inflammatory cytokines or to act as efficient antigen-presenting cells supports the theory that tissue damage may directly trigger innate inflammatory reactions.[8,56,57] This concept was hypothesized by Dr. Polly Matzinger, who referred to it as the "danger response theory."[58] Although the danger theory was originally based on little scientific evidence, there is now convincing data to indicate that cells do indeed respond to damaged tissues and cells. Some of these endogenous mediators related to the injury response that may play a role in stimulating the release of inflammatory cytokines following injury include several heat shock proteins (e.g., HSP60, HSP70, and GP96), components of the clotting cascade (fibronectin A and fibrinogen), chromatin-IgG complexes, and HMGB1.[8,59–63] These endogenous mediators of inflammation are appropriately named damage-associated molecular patterns (DAMPs) because they alert the immune system that tissue damage has occurred. Interestingly, many of these expanding lists of endogenous DAMPs have been shown to interact with TLRs (Table 10.1). Another interesting molecule that also acts as a potent DAMP is uric acid crystals (monosodium urate, or MSU).[64] Following stress and tissue damage by necrosis, MSU is released and has been shown to activate dendritic cells. Thus, there are also DAMPs that interact with non-TLR innate receptors on cells. We would predict that there are undiscovered DAMPs and DAMP receptors that play central roles in regulating the mammalian immune system response to injury and infection. The identification and characterization of these endogenous DAMPs and DAMP receptors would represent a major advance in linking injury responses to the initiation of specific changes in innate and adaptive immune function.

The primary focus of our laboratory is to identify and characterize the influence of injury on the immune system. To accomplish this, our laboratory uses a mouse injury model and also performs studies using peripheral blood cells prepared from burn and trauma patients.[65–67] Our mouse model is a 25% total body surface area (TBSA), full-thickness burn injury that causes a well-demarcated injury. The advantages of this model are that it has a mortality rate of less than 5% and that the injury damages the nerves in the skin, making it an anesthetic and humane injury model. Moreover, the mice do not need long-term analgesic treatment, which can interfere with normal regulation of immune responses. Using this model, we have found that injury causes a progressive increase in the inflammatory reactivity of innate immune cell types and a reduction in CD4 T cell–mediated immune reactivity. Studies performed by other groups using the burn model and other related mouse injury models have confirmed our observations. Most importantly, similar observations have been documented in severely injured patients. Since this chapter addresses changes in TLR response after injury, we will describe methods that we used to study the impact of injury on innate immune system responses.

10.2 STUDY OF TWO-HIT RESPONSE IN MICE

We hypothesized that injury might augment innate immune cell responses by altering TLR4 reactivity based on observations made in some early studies showing that peripheral blood mononuclear cells prepared from burn or trauma patients showed a higher cytokine response to *E. coli* endotoxin and LPS.[68,69] Interestingly, it was these findings along with clinical observations that precipitated the idea that injury induces a systemic inflammatory response that involves enhanced reactivity to bacteria and their toxins.[70,71] Subsequently, a series of clinical observations set in motion the notion that injury causes a systemic inflammatory response syndrome (SIRS) that is characterized by a number of clinical symptoms. A recent report describes and reviews the current clinical features of SIRS.[70] An additional attribute of the injury-induced SIRS response is that it primes the host for an enhanced secondary inflammatory response referred to as the "two-hit" response.[72] Because severely injured patients are predisposed to develop opportunistic infections, it is widely assumed that the two-hit response following injury promotes multiple organ failure (MOF) in trauma patients who develop infectious complications. We believe that the progressive increase in the proinflammatory state that occurs after injury is a major contributor to the development of the two-hit response phenotype after injury. Since these described features of the injury response appear to involve phenotypic changes in the level of reactivity of the innate immune system to bacteria, we wished to study how injury might influence TLR4 responses with a primary focus on the development of SIRS and the two-hit response.

We have developed a two-hit model to test the in vivo significance of injury-induced enhancement of TLR4 reactivity.[73] We found that burn-injured mice were significantly more susceptible to LPS-induced shock. The general approach for these studies is to challenge control (sham) mice versus burn mice with LPS and then measure mortality rates or immune responses as described below. The LPS dose used for this work is sublethal for sham mice but represents an LD50 for burn mice. This allows for clear outcomes in mortality studies. The in vivo cytokine responses that we typically measure include serum and organ extracts made at various time points after LPS challenge. However, additional endpoint determinations can be made (e.g., liver enzyme analyses) to study organ damage.

10.2.1 MOUSE BURN INJURY MODEL

Our mouse burn injury protocol has been approved by the National Institutes of Health and the Harvard Medical School Standing Committee on Animal Research. If performed correctly, the mortality for this injury model will be < 5% when the injury is delivered to mice larger than 22 g.

Protocol 10.1: Mouse Burn Injury Model

1. Anesthetize mice by intraperitoneal (i.p.) injection of ketamine (125 mg/kg) with xylazine (20 mg/kg).
2. Shave the dorsal fur.
3. Place mice in an insulated plastic mold to expose 25% total body surface area. A 60 cc syringe with an appropriately sized opening works well as a mold.

4. The exposed dorsum should then be immersed in 90 °C water for 9 seconds. Sham mice undergo identical treatment, but are exposed to isothermic (24 °C) water.
5. Finally, all mice are resuscitated by an i.p. injection of 1 ml of 0.9% pyrogenfree saline.

Most commercially available preparations of LPS contain contaminants that can activate TLR2 in addition to TLR4. Therefore, we use the protocol adapted from Hirschfeld et al.74 to repurify LPS, making it a more specific ligand for TLR4.

Protocol 10.2: Purification of LPS

Wear gloves, eye protection, and a mask!

1. Suspend 5 mg of LPS (Sigma-Aldrich, St. Louis, MO) in 1 ml of PBS.
2. Add deoxycholate (DOC) sodium salt (Sigma-Aldrich) to a final concentration of 0.5% w/v (i.e., by weight). It is best to make a 10× stock of DOC and add it to LPS suspended in PBS.
3. Add an equal volume of water-saturated phenol solution (Invitrogen, Carlsbad, CA).
4. Vortex samples intermittently for 5 minutes.
5. Allow phases to separate at room temperature for 5 minutes.
6. Place on ice for 5 minutes.
7. Centrifuge at 3,000 rpm for 20 minutes.
8. Transfer the top layer to a fresh tube.
9. Reextract the phenol layer with PBS and 0.5% DOC, as in steps 4–8.
10. Pool the aqueous phases, and adjust to 75% ethanol and 30 mM sodium acetate.
11. Precipitate for 1 hour at –20 °C.
12. Centrifuge at 3,000 rpm for 20 minutes.
13. Wash pellet once by adding 100% ice-cold ethanol.
14. Centrifuge at 3,000 rpm for 5 minutes.
15. Pour off ethanol, and let pellet air dry.
16. Resuspend pellet in 1 ml of PBS.

10.2.2 ORGAN AND SERUM CYTOKINE RESPONSES

Protocol 10.3: Assessment of Plasma and Organ Cytokine Levels

1. Challenge sham and burn mice by i.p. injection with repurified *E. coli* LPS. The LD50 dose for burn-injured mice is approximately 5 mg/kg, whereas

this is sublethal for sham mice. *Note*: the LD50 needs to be measured empirically for each batch of repurified LPS prior to performing two-hit studies.

2. Mortality should be followed for 7 days, since deaths rarely occur after 4 days.
3. At time 0, and at multiple time points up to 8 hours after LPS injection, collect blood into heparinized 1 ml syringes.
4. Prepare plasma by centrifugation at 3,000 × g for 20 minutes at 4 °C.
5. Following exsanguinations, harvest "pea-sized" portions of lung, liver, kidney, or spleen into 1.5 ml microcentrifuge tubes and add 0.5 ml of ice-cold PBS containing the recommended concentration of Complete protease inhibitor cocktail (Roche Applied Science, Basel, Switzerland).
6. Homogenize tissues for 30 seconds each, on ice, using a tissue-tearer device (BioSpec Products, Bartlesville, OK).
7. Clarify tissue extracts by centrifugation at 3,000 × g for 20 minutes at 4 °C.
8. Transfer clarified tissue extracts to fresh tubes.
9. Determine protein amounts in each sample by a colorimetric protein assay. We use the Bradford reagent (Sigma-Aldrich, St. Louis, MO).
10. Store plasma and organ extracts at –20 °C prior to performing cytokine and chemokine measurements.

There are a number of different types of assays that can be used to measure cytokines. We have used a combination of ELISA, cytokine bead array assay (BD Biosciences, San Jose, CA), and Luminex assays (Luminex, Austin, TX) for our studies. The assay of choice should have a sensitivity cutoff of less than 10 pg/ml cytokine if possible. The bead array assays are generally more expensive, but are easier to perform and more accurate than ELISA.

10.2.3 EX VIVO STUDIES TO MEASURE IMMUNE CELL RESPONSES

A common approach in many of our studies is to measure how spleen or lymph node cells respond to innate or adaptive immune cell stimulation. The primary readout for these studies is cytokine production. Below is a list of TLR agonists commonly used in our laboratory for stimulation:

LPS: Cell wall component of gram-negative bacteria activates cells that express TLR4 and CD14. We repurify LPS purchased from Sigma-Aldrich (St. Louis, MO) or Difco (Franklin Lakes, NJ). The usual range used for stimulation is from 10 ng/ml to 1 µg/ml.

PGN: Cell wall component of gram-positive bacteria activates cells that express TLR2. The usual range for stimulation is from 100 ng/ml to 10 µg/ml. We use certified endotoxin-free PGN from Toxin Technology (Sarasota, FL).

CpG DNA: Unmethylated bacterial DNA, 1826 ODN, or other ODNs (e.g., 2006 and 2336) from Coley Pharmaceuticals Group (Wellesley, MA). The usual dose range is from 10 ng/ml to 1 µg/ml.

BLP (or Pam$_3$Cys): Common peptide found in the cell walls of gram-negative and gram-positive bacteria. Activates cells via TLR2. The usual dose range is 10 ng/ml to 1 µg/ml. Pam$_3$Cys can be obtained from Bachem (Weil am Rhein, Germany).

Protocol 10.4: Isolation of Splenocytes

1. Mince whole spleen in 1 ml of C5 culture medium (below) with a sterile stainless steel mesh screen placed in a small sterile Petri dish and 5 ml syringe plunger.
2. Centrifuge at 300 × g for 10 minutes.
3. Resuspend cell pellet in 5 ml of red blood cell lysis buffer (below) per spleen, and incubate with gentle rocking for 5 minutes at room temperature.
4. Add 5 ml of C5 culture medium, and centrifuge at 300 × g for 10 minutes.
5. Wash cells in 10 ml of C5 culture medium, followed by centrifugation at 300 × g for 10 minutes.
6. Count cells in a hemacytometer, and seed 10^5–10^6 cells per well in a 96-well U-bottom plate in a total volume of 200 µl of C5 culture medium per well.
7. Culture cells at 37 °C under 5% CO_2 for 48 hours.

Red Cell Lysis Buffer

8.29 g of NH$_4$Cl (0.15 M)
1 g of KHCO$_3$ (10.0 mM)
37.2 mg of Na$_2$EDTA (0.1 mM)
Add 800 ml of H$_2$O, and adjust pH to 7.2–7.4 with 1 N HCl.
Add H$_2$O to 1 liter.
Filter, sterilize (0.2 µm), and store at room temperature.

C5 Culture Medium

We use RPMI-1640 supplemented with 5% heat-inactivated (56 °C for 40 minutes) fetal bovine serum (FBS) from GIBCO/Life Technologies (Carlsbad, CA). Other additives include antibiotic/antimycotic mixture, HEPES, nonessential amino acids, and glutamine, all purchased as 100× stocks from Invitrogen (Carlsbad, CA); and 2.5 µM β-mercaptoethanol.

We use intracellular cytokine staining to identify the cellular source of cytokines produced in our ex vivo TLR stimulation assays. The assay requires the use of a Golgi-blocking drug to trap the cytokines being produced inside the cells. We use brefeldin A purchased from Sigma-Aldrich (St. Louis, MO), but other groups use monensin as a Golgi-blocking drug. The following protocol is a general protocol for most cytokines that we stain in our laboratory.

Protocol 10.5: Intracellular Cytokine Staining

1. Aliquot 10^6 cells per well in a round-bottom 96-well plate.
2. Stimulate cells for 6 hours at 37 °C in the presence of brefeldin A added at a final concentration of 10 μg/ml.
3. After 6 hours, spin down plates and resuspend pellets in 20 μl of FC block reagent (antibody cocktail specific for Fc receptors; BD Biosciences, San Jose, CA) diluted to 1 μg/ml in PBA (1% BSA and 0.1% sodium azide in PBS).
4. Incubate at room temperature for 10 minutes.
5. Add 20 μl of fluorescently labeled antibody specific for the desired cell marker at a concentration of 1 μg/ml in PBA. For example, we use anti-F4/80 antibody to identify macrophages. These antibodies can be purchased from a number of sources: BD Biosciences, eBioscience (San Diego, CA), BioLegend (San Diego, CA), or CALTAG (Carlsbad, CA).
6. After 20 minutes of incubation in the dark at 4 °C, spin down plates.
7. Resuspend cells in 50 μl of Fix buffer (BioLegend), and incubate in the dark at 4 °C for 20 minutes.
8. Spin down plates and resuspend in 50 μl of Perm buffer (BioLegend) containing 1 μg/ml of normal rat IgG and mouse IgG (CALTAG), then incubate for 10 minutes in the dark at 4 °C.
9. Add 20 μl of the desired fluorescently labeled anticytokine antibody diluted to 2 μg/ml in Perm buffer.
10. Incubate for 30 minutes in the dark at 4 °C.
11. Spin down plates, and fix cells in 0.3% paraformaldehyde in PBS.
12. Analyze samples by flow cytometry to determine the percentage of positive-staining cells as well as the mean fluorescence intensity (MFI).

10.2.4 ASSESSING TLR4 SIGNALING IN WHOLE CELLS

With the availability of antibodies specific for the phosphorylated domains of many signaling molecules, it is now feasible to study cell signaling in whole cells by examining changes in the level of phosphorylated molecules by flow cytometry rather than by Western immunoblot procedures. A general approach for these studies is to stimulate cells and then prepare them for flow cytometry by fixation and permeabilization. Fluorescently tagged antibodies specific for the phosphorylated portion, usually antiphosphopeptide specific, are readily available for these studies (Cell Signaling Technologies, Beverly, MA). In our work, we used this approach to compare which TLR4-stimulated MAPK-signaling molecule was more or less phosphorylated in macrophages from sham versus burn mice. We found the p38 was more heavily phosphorylated in LPS-stimulated macrophages from burn as compared to sham mice.[54] We combined our flow cytometry approach with using pharmacologic inhibitors of p38 signaling to prove that our findings were specific to TLR4-stimulated p38 activation in macrophages.

Protocol 10.6: Monitoring TLR4 Signaling by Flow Cytometry

1. Add 50 μl of cell suspensions (5×10^5 cell/ml) per well in 96-well plates. Incubate for 2 hours at 37 °C to achieve steady state for signaling study.
2. Add 50 μl of a TLR agonist or 50 μl of culture medium for control.
3. Incubate at 37 °C for 5, 15, 30, and 60 minutes, respectively.
4. Fix cells by adding 100 μl of 1.5% paraformaldehyde in PBS and incubating at 37 °C for 10 minutes.
5. Spin at $300 \times g$ for 10 minutes, and vortex.
6. Resuspend cells in 200 μl of ice-cold methanol for 10 minutes. Samples can be stored in methanol at –20 °C.
7. Wash with 200 μl of PBA prior to performing the intracellular staining procedure.
8. Block FC receptor. Add 20 μl of FC block diluted 1:50 in PBA, and incubate at room temperature for 15 minutes.
9. Add 20 μl of primary antibody (1:50 dilution): rabbit Ig control, p38 MAPK, pERK, and pSAPK/JNK, and phospho-p65 NF-κB or antibodies specific for nonphosphorylated p38, ERK1/2, and SAPK/JNK. To detect macrophages, surface-stain for F4/80 by adding 20 μl of antibody (1 _μg/ml) in PBA.
10. Incubate at room temperature for 30 minutes in the dark.
11. Wash cells with 200 μl of PBA followed by centrifugation at $300 \times g$ for 10 minutes.
12. Add 20 μl of PE-conjugated secondary goat-anti-rabbit antibody (diluted to 1:100), and incubate again for 30 minutes at room temperature in the dark.
13. Wash once in PBA.
14. Resuspend in 200 μl of 0.3% paraformaldehyde in PBS.
15. Determine the percentage of positive-staining cells as well as the MFI by flow cytometry and FACS analysis.

10.3 CONCLUDING REMARKS

This chapter provides a brief overview of some of the techniques available to study TLR responses in vivo and in vitro. The primary objectives of our studies relevant to TLR biology are to determine how injury alters the reactivity of innate immune cells to bacterial antigens or toxins with the overall aim to understand the basic mechanisms associated with trauma-induced immunological complications. We are also interested in the discovery of endogenous DAMPs that either act through TLRs or modulate the response of innate immune cells to TLR stimuli. The identification of endogenous molecules important in stimulating the systemic inflammatory response to major injury will likely provide new insights into therapeutic approaches to benefit severely injured patients. Moreover, the approaches described here are immediately applicable to human studies with the limitation of using peripheral blood cells to study TLR responses. We have compared many of the responses seen in injured mice to those occurring in severely injured patients, and our collective findings indicate that most of the changes observed in mice also occur in humans.[66,67,69,75] We believe this validates and justifies our ongoing research studies using mice to model the mammalian injury response. We hope that the description of these protocols will be of benefit to your own research efforts.

ACKNOWLEDGMENTS

This work was supported by NIH NIGMS Grants RO1GM57664 and RO135633. Further support was provided by the Brook Family and the Julian and Eunice Research Funds.

REFERENCES

1. Hashimoto, C., Hudson, K. L., and Anderson, K. V. The Toll gene of Drosophila, required for dorsal-ventral embryonic polarity, appears to encode a transmembrane protein. Cell 52 (2), 269–79, 1988.
2. Anderson, K. V., Jurgens, G., and Nusslein-Volhard, C. Establishment of dorsal-ventral polarity in the Drosophila embryo: genetic studies on the role of the Toll gene product. Cell 42 (3), 779–89, 1985.
3. Lemaitre, B., Nicolas, E., Michaut, L., Reichhart, J. M., and Hoffmann, J. A. The dorsoventral regulatory gene cassette spatzle/Toll/cactus controls the potent antifungal response in Drosophila adults. Cell 86 (6), 973–83, 1996.
4. Beutler, B., and Rehli, M. Evolution of the TIR, Tolls and TLRs: functional inferences from computational biology. Curr Top Microbiol Immunol 270, 1–21, 2002.
5. Rock, F. L., Hardiman, G., Timans, J. C., Kastelein, R. A., and Bazan, J. F. A family of human receptors structurally related to Drosophila Toll. Proc Natl Acad Sci USA 95 (2), 588–93, 1998.
6. Martin, M. U., and Wesche, H. Summary and comparison of the signaling mechanisms of the Toll/interleukin-1 receptor family. Biochim Biophys Acta 1592 (3), 265–80, 2002.
7. Medzhitov, R., and Janeway, C., Jr. Innate immune recognition: mechanisms and pathways. Immunol Rev 173, 89–97, 2000.
8. Beg, A. A. Endogenous ligands of Toll-like receptors: implications for regulating inflammatory and immune responses. Trends Immunol 23 (11), 509–12, 2002.
9. Biragyn, A., Ruffini, P. A., Leifer, C. A., Klyushnenkova, E., Shakhov, A., Chertov, O., Shirakawa, A. K., Farber, J. M., Segal, D. M., Oppenheim, J. J., and Kwak, L. W. Toll-like receptor 4-dependent activation of dendritic cells by beta-defensin 2. Science 298 (5595), 1025–9, 2002.
10. El Mezayen, R., El Gazzar, M., Seeds, M. C., McCall, C. E., Dreskin, S. C., and Nicolls, M. R. Endogenous signals released from necrotic cells augment inflammatory responses to bacterial endotoxin. Immunol Lett 111 (1), 36–44, 2007.
11. Yu, M., Wang, H., Ding, A., Golenbock, D. T., Latz, E., Czura, C. J., Fenton, M. J., Tracey, K. J., and Yang, H. HMGB1 signals through Toll-like receptor (TLR) 4 and TLR2. Shock 26 (2), 174–79, 2006.

12. Park, J. S., Gamboni-Robertson, F., He, Q., Svetkauskaite, D., Kim, J. Y., Strassheim, D., Sohn, J. W., Yamada, S., Maruyama, I., Banerjee, A., Ishizaka, A., and Abraham, E. High mobility group box 1 protein interacts with multiple Toll-like receptors. Am J Physiol Cell Physiol 290 (3), C917–24, 2006.

13. Seong, S. Y., and Matzinger, P. Hydrophobicity: an ancient damage-associated molecular pattern that initiates innate immune responses. Nat Rev Immunol 4 (6), 469–78, 2004.

14. Oppenheim, J. J., Tewary, P., de la Rosa, G., and Yang, D. Alarmins initiate host defense. Adv Exp Med Biol 601, 185–94, 2007.

15. Bianchi, M. E. DAMPs, PAMPs and alarmins: all we need to know about danger. J Leukoc Biol 81 (1), 1–5, 2007.

16. Oppenheim, J. J., and Yang, D. Alarmins: chemotactic activators of immune responses. Curr Opin Immunol 17 (4), 359–65, 2005.

17. Poltorak, A., He, X., Smirnova, I., Liu, M. Y., Huffel, C. V., Du, X., Birdwell, D., Alejos, E., Silva, M., Galanos, C., Freudenberg, M., Ricciardi-Castagnoli, P., Layton, B., and Beutler, B. Defective LPS signaling in C3H/HeJ and C57BL/10ScCr mice: mutations in Tlr4 gene. Science 282 (5396), 2085–8, 1998.

18. Hoshino, K., Takeuchi, O., Kawai, T., Sanjo, H., Ogawa, T., Takeda, Y., Takeda, K., and Akira, S. Cutting edge: Toll-like receptor 4 (TLR4)-deficient mice are hyporesponsive to lipopolysaccharide: evidence for TLR4 as the LPS gene product. J Immunol 162 (7), 3749–52, 1999.

19. Medzhitov, R., Preston-Hurlburt, P., and Janeway, C. A., Jr. A human homologue of the Drosophila Toll protein signals activation of adaptive immunity. Nature 388 (6640), 394–97, 1997.

20. Takeda, K., Kaisho, T., and Akira, S. Toll-like receptors. Annu Rev Immunol 9, 9, 2003.

21. Takeda, K., and Akira, S. Toll-like receptors in innate immunity. Int Immunol 17 (1), 1–14, 2005.

22. Alexopoulou, L., Holt, A, C., Medzhitov, R., and Flavell, R. A. Recognition of double-stranded RNA and activation of NF-kappaB by Toll-like receptor 3. Nature 413 (6857), 732–78, 2001.

23. Krieg, A. M. CpG motifs in bacterial DNA and their immune effects. Annu Rev Immunol 20, 709–60, 2002.

24. Boonstra, A., Asselin-Paturel, C., Gilliet, M., Crain, C., Trinchieri, G., Liu, Y. J., and O'Garra, A. Flexibility of mouse classical and plasmacytoid-derived dendritic cells in directing T helper type 1 and 2 cell development: dependency on antigen dose and differential Toll-like receptor ligation. J Exp Med 197 (1), 101–9, 2003.

25. Dabbagh, K., and Lewis, D. B. Toll-like receptors and T-helper-1/T-helper-2 responses. Curr Opin Infect Dis 16 (3), 199–204, 2003.

26. Schwarz, K., Storni, T., Manolova, V., Didierlaurent, A., Sirard, J. C., Rothlisberger, P., and Bachmann, M. F. Role of Toll-like receptors in costimulating cytotoxic T cell responses. Eur J Immunol 33 (6), 1465–70, 2003.

27. Pasare, C., and Medzhitov, R. Toll-like receptors: linking innate and adaptive immunity. Microbes Infect 6 (15), 1382–7, 2004.

28. Yamamoto, M., Sato, S., Mori, K., Hoshino, K., Takeuchi, O., Takeda, K., and Akira, S. Cutting edge: a novel Toll/IL-1 receptor domain-containing adapter that preferentially activates the IFN-beta promoter in the Toll-like receptor signaling. J Immunol 169 (12), 6668–72, 2002.

29. Trinchieri, G. Interleukin-12 and the regulation of innate resistance and adaptive immunity. Nat Rev Immunol 3 (2), 133–46, 2003.

30. Yamamoto, M., Sato, S., Hemmi, H., Hoshino, K., Kaisho, T., Sanjo, H., Takeuchi, O., Sugiyama, M., Okabe, M., Takeda, K., and Akira, S. Role of adaptor TRIF in the MyD88-independent Toll-like receptor signaling pathway. Science 301 (5633), 640–43, 2003.
31. Kawai, T., Sato, S., Ishii, K. J., Coban, C., Hemmi, H., Yamamoto, M., Terai, K., Matsuda, M., Inoue, J., Uematsu, S., Takeuchi, O., and Akira, S. Interferon-alpha induction through Toll-like receptors involves a direct interaction of IRF7 with MyD88 and TRAF6. Nat Immunol 5 (10), 1061–8, 2004.
32. Kawai, T., Adachi, O., Ogawa, T., Takeda, K., and Akira, S. Unresponsiveness of MyD88-deficient mice to endotoxin. Immunity 11 (1), 115–22, 1999.
33. Takeda, K., and Akira, S. TLR signaling pathways. Semin Immunol 16 (1), 3–9, 2004.
34. Yamamoto, M., Sato, S., Hemmi, H., Sanjo, H., Uematsu, S., Kaisho, T., Hoshino, K., Takeuchi, O., Kobayashi, M., Fujita, T., Takeda, K., and Akira, S. Essential role for TIRAP in activation of the signalling cascade shared by TLR2 and TLR4. Nature 420 (6913), 324–29, 2002.
35. Horng, T., Barton, G. M., and Medzhitov, R. TIRAP: an adapter molecule in the Toll signaling pathway. Nat Immunol 2 (9), 835–41, 2001.
36. Latz, E., Visintin, A., Lien, E., Fitzgerald, K. A., Monks, B. G., Kurt-Jones, E. A., Golenbock, D. T., and Espevik, T. Lipopolysaccharide rapidly traffics to and from the Golgi apparatus with the Toll-like receptor 4-MD-2-CD14 complex in a process that is distinct from the initiation of signal transduction. J Biol Chem 277 (49), 47834–43, 2002.
37. Espevik, T., Latz, E., Lien, E., Monks, B., and Golenbock, D. T. Cell distributions and functions of Toll-like receptor 4 studied by fluorescent gene constructs. Scand J Infect Dis 35 (9), 660–64, 2003.
38. Triantafilou, M., Gamper, F. G., Haston, R. M., Mouratis, M. A., Morath, S., Hartung, T., and Triantafilou, K. Membrane sorting of Toll-like receptor (TLR)-2/6 and TLR2/1 heterodimers at the cell surface determines heterotypic associations with CD36 and intracellular targeting. J Biol Chem 281 (41), 31002–11, 2006.
39. Soong, G., Reddy, B., Sokol, S., Adamo, R., and Prince, A. TLR2 is mobilized into an apical lipid raft receptor complex to signal infection in airway epithelial cells. J Clin Invest 113 (10), 1482–9, 2004.
40. Triantafilou, M., and Triantafilou, K. Receptor cluster formation during activation by bacterial products. J Endotoxin Res 9 (5), 331–35, 2003.
41. Pfeiffer, A., Bottcher, A., Orso, E., Kapinsky, M., Nagy, P., Bodnar, A., Spreitzer, I., Liebisch, G., Drobnik, W., Gempel, K., Horn, M., Holmer, S., Hartung, T., Multhoff, G., Schutz, G., Schindler, H., Ulmer, A. J., Heine, H., Stelter, F., Schutt, C., Rothe, G., Szollosi, J., Damjanovich, S., and Schmitz, G. Lipopolysaccharide and ceramide docking to CD14 provokes ligand-specific receptor clustering in rafts. Eur J Immunol 31 (11), 3153–64, 2001.
42. Suzuki, N., Suzuki, S., Duncan, G. S., Millar, D. G., Wada, T., Mirtsos, C., Takada, H., Wakeham, A., Itie, A., Li, S., Penninger, J. M., Wesche, H., Ohashi, P. S., Mak, T. W., and Yeh, W. C. Severe impairment of interleukin-1 and Toll-like receptor signalling in mice lacking IRAK-4. Nature 416 (6882), 750–56, 2002.
43. Li, S., Strelow, A., Fontana, E. J., and Wesche, H. IRAK-4: a novel member of the IRAK family with the properties of an IRAK-kinase. Proc Natl Acad Sci USA 99 (8), 5567–72, 2002.
44. O'Neill, L. A., and Bowie, A. G. The family of five: TIR-domain-containing adaptors in Toll-like receptor signalling. Nat Rev Immunol 7 (5), 353–64, 2007.
45. Symons, A., Beinke, S., and Ley, S. C. MAP kinase kinase kinases and innate immunity. Trends Immunol 27 (1), 40–48, 2006.

46. Doyle, S. E., O'Connell, R., Vaidya, S. A., Chow, E. K., Yee, K., and Cheng, G. Toll-like receptor 3 mediates a more potent antiviral response than Toll-like receptor 4. J Immunol 170 (7), 3565–71, 2003.

47. Akira, S., and Hoshino, K. Myeloid differentiation factor 88-dependent and -independent pathways in Toll-like receptor signaling. J Infect Dis 187 (Suppl. 2), S356–63, 2003.

48. Fitzgerald, K. A., McWhirter, S. M., Faia, K. L., Rowe, D. C., Latz, E., Golenbock, D. T., Coyle, A. J., Liao, S. M., and Maniatis, T. IKKepsilon and TBK1 are essential components of the IRF3 signaling pathway. Nat Immunol 4 (5), 491–96, 2003.

49. Leadbetter, E. A., Rifkin, I. R., Hohlbaum, A. M., Beaudette, B. C., Shlomchik, M. J., and Marshak-Rothstein, A. Chromatin-IgG complexes activate B cells by dual engagement of IgM and Toll-like receptors. Nature 416 (6881), 603–7, 2002.

50. Paterson, H. M., Murphy, T. J., Purcell, E. J., Shelley, O., Kriynovich, S. J., Lien, E., Mannick, J. A., and Lederer, J. A. Injury primes the innate immune system for enhanced Toll-like receptor reactivity. J Immunol 171 (3), 1473–83, 2003.

51. Fan, H., and Cook, J. A. Molecular mechanisms of endotoxin tolerance. J Endotoxin Res 10 (2), 71–84, 2004.

52. Medvedev, A. E., Kopydlowski, K. M., and Vogel, S. N. Inhibition of lipopolysaccharide-induced signal transduction in endotoxin-tolerized mouse macrophages: dysregulation of cytokine, chemokine, and Toll-like receptor 2 and 4 gene expression. J Immunol 164 (11), 5564–74, 2000.

53. West, M. A., and Heagy, W. Endotoxin tolerance: a review. Crit Care Med 30 (1 Suppl.), S64–73, 2002.

54. Maung, A. A., Fujimi, S., Miller, M. L., Macconmara, M. P., Mannick, J. A., and Lederer, J. A. Enhanced TLR4 reactivity following injury is mediated by increased p38 activation. J Leukoc Biol 78, 565–73, 2005.

55. Medvedev, A. E., Piao, W., Shoenfelt, J., Rhee, S. H., Chen, H., Basu, S., Wahl, L. M., Fenton, M. J., and Vogel, S. N. Role of TLR4 tyrosine phosphorylation in signal transduction and endotoxin tolerance. J Biol Chem 282 (22), 16042–53, 2007.

56. Li, M., Carpio, D. F., Zheng, Y., Bruzzo, P., Singh, V., Ouaaz, F., Medzhitov, R. M., and Beg, A. A. An essential role of the NF-kappa B/Toll-like receptor pathway in induction of inflammatory and tissue repair gene expression by necrotic cells. J Immunol 166 (12), 7128–35, 2001.

57. Gallucci, S., Lolkema, M., and Matzinger, P. Natural adjuvants: endogenous activators of dendritic cells. Nat Med 5 (11), 1249–55, 1999.

58. Matzinger, P. Tolerance, danger, and the extended family. Annu Rev Immunol 12, 991–1045, 1994.

59. Vabulas, R. M., Ahmad-Nejad, P., Ghose, S., Kirschning, C. J., Issels, R. D., and Wagner, H. HSP70 as endogenous stimulus of the Toll/interleukin-1 receptor signal pathway. J Biol Chem 277 (17), 15107–12, 2002.

60. Habich, C., Baumgart, K., Kolb, H., and Burkart, V. The receptor for heat shock protein 60 on macrophages is saturable, specific, and distinct from receptors for other heat shock proteins. J Immunol 168 (2), 569–76, 2002.

61. Vabulas, R. M., Ahmad-Nejad, P., da Costa, C., Miethke, T., Kirschning, C. J., Hacker, H., and Wagner, H. Endocytosed HSP60s use Toll-like receptor 2 (TLR2) and TLR4 to activate the Toll/interleukin-1 receptor signaling pathway in innate immune cells. J Biol Chem 276 (33), 31332–9, 2001.

62. Ohashi, K., Burkart, V., Flohe, S., and Kolb, H. Cutting edge: heat shock protein 60 is a putative endogenous ligand of the Toll-like receptor-4 complex. J Immunol 164 (2), 558–61, 2000.

63. Tsung, A., Klune, J. R., Zhang, X., Jeyabalan, G., Cao, Z., Peng, X., Stolz, D. B., Geller, D. A., Rosengart, M. R., and Billiar, T. R. HMGB1 release induced by liver ischemia involves Toll-like receptor 4 dependent reactive oxygen species production and calcium-mediated signaling. J Exp Med 204 (12), 2913–23, 2007.
64. Shi, Y., Evans, J. E., and Rock, K. L. Molecular identification of a danger signal that alerts the immune system to dying cells. Nature 425 (6957), 516–21, 2003.
65. Purcell, E. M., Dolan, S. M., Kriynovich, S., Mannick, J. A., and Lederer, J. A. Burn injury induces an early activation response by lymph node cd4+ T cells. Shock 25 (2), 135–40, 2006.
66. MacConmara, M. P., Maung, A. A., Fujimi, S., McKenna, A. M., Delisle, A., Lapchak, P. H., Rogers, S., Lederer, J. A., and Mannick, J. A. Increased CD4+ CD25+ T regulatory cell activity in trauma patients depresses protective Th1 immunity. Ann Surg 244 (4), 514–23, 2006.
67. Murphy, T., Paterson, H., Rogers, S., Mannick, J. A., and Lederer, J. A. Use of intracellular cytokine staining and bacterial superantigen to document suppression of the adaptive immune system in injured patients. Ann Surg 238 (3), 401–10, discussion 410–11, 2003.
68. Menger, M. D., and Vollmar, B. Surgical trauma: hyperinflammation versus immunosuppression? Langenbecks Arch Surg 389 (6), 475–84, 2004.
69. Molloy, R. G., O'Riordain, M., Holzheimer, R., Nestor, M., Collins, K., Mannick, J. A., and Rodrick, M. L. Mechanism of increased tumor necrosis factor production after thermal injury: altered sensitivity to PGE2 and immunomodulation with indomethacin. J Immunol 151 (4), 2142–49, 1993.
70. Levy, M. M., Fink, M. P., Marshall, J. C., Abraham, E., Angus, D., Cook, D., Cohen, J., Opal, S. M., Vincent, J. L., and Ramsay, G. 2001 SCCM/ESICM/ACCP/ATS/SIS International Sepsis Definitions Conference. Intensive Care Med 29 (4), 530–38, 2003.
71. Afessa, B., Green, B., Delke, I., and Koch, K. Systemic inflammatory response syndrome, organ failure, and outcome in critically ill obstetric patients treated in an ICU. Chest 120 (4), 1271–7, 2001.
72. Moore, F. A., Moore, E. E., and Read, R. A. Postinjury multiple organ failure: role of extrathoracic injury and sepsis in adult respiratory distress syndrome. New Horiz 1 (4), 538–49, 1993.
73. Murphy, T. J., Paterson, H. M., Kriynovich, S., Zang, Y., Kurt-Jones, E. A., Mannick, J. A., and Lederer, J. A. Linking the "two-hit" response following injury to enhanced TLR4 reactivity. J Leukoc Biol 77 (1), 16–23, 2005.
74. Hirschfeld, M., Ma, Y., Weis, J. H., Vogel, S. N., and Weis, J. J. Cutting edge: repurification of lipopolysaccharide eliminates signaling through both human and murine toll-like receptor 2. J Immunol 165 (2), 618–22, 2000.
75. Goebel, A., Kavanagh, E., Lyons, A., Saporoschetz, I. B., Soberg, C., Lederer, J. A., Mannick, J. A., and Rodrick, M. L. Injury induces deficient interleukin-12 production, but interleukin-12 therapy after injury restores resistance to infection. Ann Surg 231 (2), 253–61, 2000.

Index

Note: locators for figures are in italics, for tables in bold, and for protocols in bold italics

A

adapters
  Mal (MyD88 adaptor-like), 46
  MyD88 (myeloid differentiation factor 88), 46
  TRAM (TRIF-related adaptor molecules), 46
  Trif, 46
alarmin, danger signals of endogenous origin and, 25
Alzheimer's disease, 152
antibodies
  affinity, 50
  anti-TLR, 24
  anti-TLR2, 45
  labeling with fluorescent dyes, *63–64*
  production of, 40
  specificity, 50
antigen presentation, 40
antigen-presenting cells (APCs), dendritic cells (DCs) and, 165
antigen-specific adaptive immunity, Toll-like receptors (TLRs) and, 2
astrocytes, 147
  isolation of from the neonatal mouse brain, *151*
  response to CNS infection, 147
astrocytes, functional gap junction channels in demonstrated by single-cell microinjection technique with Lucifer yellow, *153*
atherosclerosis, 40
autoimmune diseases, 59
  encephalomyelitis (EAE), 152
  lupus erthematosus, 110
  TLRs (Toll-like receptors) and, 23

B

bacteria
  cell walls of, 25
  Toll-like receptors (TLRs) and, 39
bacteria, phagocytosis of
  dendritic cells (DCs) and, 3
  macrophages and, 3
bacterial infection, 2–3, 44

Toll-like receptors (TLRs) and, 2–3
bacterial nucleic acids, TLR family in humans and, 59
biochemical fractionation protocol, 46
BioPORTER®, 81–82
  changes in RhoGTPase protein expression or activity, 81
BioPORTER® reagent, transduction of human neutrophils with RhoGTPases using, *81–82*
β-lactamase complementation techniques
  TLR TLR dimerization, 90
  in vivo interaction of Toll-like receptors (TLRs) and signaling adaptors, 90
β-lactamase fragments *Bla(a)*
  cloning of, *91–93*
β-lactamase fragments *Bla(a)*, cloning of, 91–92
β-lactamase fragments *Bla(b)*, *91–93*
β-lactamase fragments *Bla(b)*, cloning of, 91–92
β-lactamase PCAs, 99–105
  complementation analysis by flow cytometry, 102–104
  complementation analysis by fluorescence microscopy, 100–101
  preparation of HEK293 cells or HeLa cells for transient transfection, 99
BLP (bacterial lipopeptide), 173
BMDM (bone marrow-derived macrophage), efficiency of infection in, *121*
BMDM (bone marrow-derived macrophages)
  from mice, preparation of, *118–119*
  culture media for BMDM (500 ml), *119*
  red blood cell lysis buffer (500 ml), *118–119*
bone marrow-derived macrophages, *see* BMDM (bone marrow-derived macrophage)
brain absesses, 146
  causes of, 146
  cerebral edema and, 146
  cerebritis, 146
  computer tomography (CT), 146
  glial gap junction communication (GJC) and, 152
  infection and, 146
  magnetic resonance imagery (MRI), 146
  necrosis, 146
  Streptococcus aureus, 146
  tissue necrosis and, 147

T - #0429 - 071024 - C2 - 234/156/10 - PB - 9780367387181 - Gloss Lamination